PETROFORMS

PETROFORMS

*Oil and the Shaping of
Nigerian Aesthetics*

HELEN KAPSTEIN

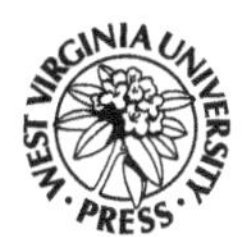

WEST VIRGINIA UNIVERSITY PRESS

MORGANTOWN

CONTENTS

ILLUSTRATIONS

ACKNOWLEDGMENTS

I believe that we scholars end up theorizing about the personal, even if it seems far removed or farfetched. I think of my own grandfather, to whom this book is dedicated: When he analyzes Shelley's "Mont Blanc" for the tensions between freedom and necessity (Kapstein, I. J.), is it too much to think about how he, as a young man, had to give his father the money he had saved for college by working on the railway, delaying freedom for necessity (Brodsky 308)? We read ourselves into our work and our work into ourselves. This book is about form and formlessness, containment and excess, control and resistance. I have been interested in pushing these boundaries from the beginning—just ask my parents.

So, my first thanks go to them—Jonathan and Nancy—for their pride and encouragement. Also to my children, Daphne and Angus, for their love and enthusiasm (and for their expertise in things beyond my ken); to Adrienne for being the best sister; and to my husband, Paul, for his support, patience, and insight as my first reader of everything. I love you all.

Friends like Melissa Amdur, Dohra Ahmad, and Lara Shapiro have always cheered me on. Debbie would have been the first person to buy this book (and she would have made me sign it). Imre Szeman, Stacey Balkan, Michael Malouf, and other comrades from the *Oil Fictions* community have been great boosters. Readers, reviewers, and editors of both this book and related publications have given me much to think about and pushed me toward bolder claims and better work. I have squeezed every last drop of insight out of my reader reports and put it to good use. Everyone at West Virginia University Press has been super (especially considering the trying circumstances of the last couple of years), including those who have moved on, like my original acquiring editor, Derek Krissoff. Friends and colleagues in my home institution, John Jay College, share genuine collegiality, something I do not take for granted. Here's to my fellow English Department professors. The Cultural Studies Association (CSA) has given me intellectual and social sustenance for many years—cheers to everyone in that orbit. I feel comforted to know I can always count on mentors and old classmates, including Rob Nixon, Gauri Viswanathan, Tawia Ansah, Claudia Stokes, Mario Ortiz-Robles, and Guillermina De Ferrari. I wish Roger Henkle could see me now.

Many people were generous with their time and thoughtful feedback (from answering a stranger's email to making a mood board for my cover art). I appreciate Sokari Douglas Camp for letting me interview her at length for the sculpture chapter and the other filmmakers and artists who gave me permission to reproduce their work. I took many notes during audience Q&As at CSA conferences, Modern Language Association meetings, and other venues where I presented some of these pages. Thank you for listening. I am grateful to Edward Powell for promoting the earlier version of my "Petrofiction" chapter (published in *Postcolonial Text*) in Oxford University Press's 2017 *The Year's Work in Critical and Cultural Theory*.

Support for this project came from multiple Professional Staff Congress—City University of New York (PSC-CUNY) grants—much appreciation both to the Research Foundation of CUNY and to the evaluators who approved my grant requests. I am also grateful to John Jay College's Office for the Advancement of Research for my Book Publication Funding Award.

Last, but far from least, thank you to librarians everywhere, but especially to those in our Lloyd Sealy Library at John Jay, who have fielded countless interlibrary loan requests and made the research for this book possible.

PETROFORMS

INTRODUCTION

"I had discovered that the plainest house can crown a fantasy or daydream. An open window can be tolerated. So can an open door. But I discovered the value of four walls and a roof. Something about containment that at the same time offers escape."

—Lloyd Jones, *Mister Pip*

In this book, I explore the riddle of "containment that at the same time offers escape." When aesthetic forms encounter the oil experience, I believe they variously lose their structure, reassemble their infrastructures, and change shape to accommodate oil's violence, disruption, and instability. These new forms try and fail to contain the excess of oil, and in that productive failure create *petroforms*.

This text, *Petroforms*, contributes a theory of form and genre to the cutting-edge field of petrocriticism, itself an offshoot of developments in postcolonial ecocriticism. Studies of resource fiction and inquiries in the energy humanities have recently taken their rightful place as necessary reflections on the role of the literary imagination in intervening in the Anthropocene as we find ourselves precariously placed in the face of a fossil fuel crisis, climate change, and mass extinction.

In *Petroforms*, I study a range of aesthetic forms in order to argue that the demands of paying attention to petroleum extraction, production, consumption, and distribution in the creation of resource fictions must necessarily alter and affect conventional forms and structures. What results is a new set of genre-bending forms (like documentary films that we can read as horror) responding to the forceful and fluid demands of the petroleum industry and its master narrative.

I take Nigeria as my geographic focus in order to craft a site-specific theory that reacts to the realities of petromodernity (the petrol world we all live in, to paraphrase Stephanie LeMenager) in one of the world's largest oil-producing nations. Production is concentrated there in the Niger Delta, resulting in a local environment steadily degraded by oil spills, flares, pollution, and contamination and a local culture shaped by false scarcity in a space of abundance, murderous politics, and escalating violence. At the same time the Niger Delta grounds my argument in its own political realities and cultural responses, Nigeria's participation in a global economy of petrodependency allows my theory of petroforms to be extended and applied more generally. Others, following on my propositions, may well find ways in which form elsewhere is also shaped and reshaped by its encounters with oil. I cover six art forms here—the short story, the romance novel, sculpture, documentary film, feature films, and drama. By no means exhaustive, this list leaves room for opportunity to do more with petropoetry (see Sule Egya) or petromusic (see Ogaga Okuyade on hip-hop), or to ask about petrophotography or petrodance.

Although each chapter of *Petroforms* tackles a different art form, read together, multiple, flexible, combining forms might actually be a full-frontal attack on or defense against Big Oil's master narrative about oil production: that the commodity belongs to the multinational corporation in conjunction with the state, and that local attempts at resource control amount to stealing or sabotage. This narrative takes its own very particular, even peculiar forms, such as the briefing notes on the Shell Nigeria website entitled "Security, Theft, Sabotage, and Spills." In the introduction to their new collection, the editors of *Containment: Technologies of Holding, Filtering, Leaking*, talk about how "leakiness . . . can be a deliberate design feature." They give oil pipelines as an example of "inevitably leaky containment," that is "a constitutive and essential—though often hidden or denied—operating feature of systems and protocols that are ostensibly designed to contain and hold". It is entirely sensible that this should be "worrying" with regard to a pipeline (Angerer et al. 9); it implies that both the physical infrastructure and its reinforcing ideological apparatuses are designed not only to permit ecologically harmful leakage, but to deflect the blame for those leaks elsewhere. When applied to the idea of aesthetic form, though, the idea of deliberate decontainment is less worrying and more liberating: Unmaking the short story or the feature film (ironically) lets it better capture how oil works.

What Peter Hitchcock calls "oil's generative law" is that it is "everywhere and obvious, it must be opaque or otherwise fantastic" (87). What he means by this is the coexistence of oil's taken-for-granted qualities, as something substrative of every level of everyday life, with its spectacular displays—gushers, spills, explosions. The ubiquity of petroleum requires that we consider not just petrofiction—as has been the standard practice of closely reading texts in the energy humanities for how they represent our engagements with oil, the automobile, the road, the pipeline, and so on—but that we recognize petroforms across the board. Oil's nature, the fact that it is everywhere, unctuously oozing into every corner of everyday life, means that it constantly spills over and out of our existing forms, genres, and systems, demanding accommodation. In order to try to contain it, we create new forms. Thus, a petroform is simultaneously reactionary (a necessary response to oil's pressures, a byproduct of the commodity itself) and resistant (an attempt at containment, at generating a retort to the very thing that shapes it). Each form figures oil and then configures and reconfigures itself in reaction to it.

At the Modern Language Association meeting in January 2020, I was struck by the number of exchanges having to do with form. It was very apparent that we, as a scholarly community, were taking a formal turn, and that there is, concurrently, an ecocritical turn to form. Even as I put the finishing touches on this book, complimentary titles appear, such as Brian Jacobson's *The Cinema of Extractions: Film Materials and Their Forms* (Columbia University Press, 2024).

But form as an idea is itself slippery, like oil. It bleeds into genre; it bleeds into content. It's hard to find a discussion of form that doesn't use one term to define the other. Inevitably, it happens here too: When I talk about Nigerian romance novels in Chapter 2, romance is a genre, with generic conventions, but the romance novel is also a specific form (itself a subset of the novel form) with its own specific formulae. In the introduction to her book *The Order of Forms* (2019), Anna Kornbluh embraces form as a project of building, contrasting herself with Caroline Levine who, in *Forms: Whole, Rhythm, Hierarchy, Network* (2017), explores the potential of "escaping from enclosing shapes" (150). I position myself between these two poles, or rather petroforms necessitate that I do; petroleum generates forms that are in excess of enclosed shapes and then, in forcing new forms, builds new possibilities. One way to think about this is to say that oil, as content, produces its own container. The ontology of oil—economic, political, and otherwise—is that it wants to saturate, to penetrate, to be ubiquitous, but it must also be contained in order to be put to work, to turn from substance into commodity. Because of its tendency to spill and seep, it regularly overflows its containers, whether they are pipelines sabotaged by desperate locals, or received art forms, like the documentary film or the short story. In so doing, oil transforms landscapes and relations and generates new forms of expression.

Both Levine and Kornbluh value form as a political device, although they arrive there by different means. Both disdain formlessness—Levine uses the word once, to say "formless or antiformal experiences have actually drawn *too much* attention from literary and cultural critics in the past few decades" (9),[1] and it destabilizes Kornbluh's core claim that form offers possibility, not constraint. *Petroforms* reclaims the politics of formlessness, traveling in the company of Georges Bataille's "Critical Dictionary" from 1929, in which the entry "formless" pairs with the entry "spittle": "*Formless* is not only an adjective with a given meaning but a term which declassifies. . . . To declare . . . that the universe is not like anything, and is simply *formless*, is tantamount to saying the universe is something like a spider or spittle" (Faccini 27). Produced in the "oily factory of mastication" (Faccini 30), "spittle is finally, through its inconsistency, its indefinite contours, the relative imprecision of its color, and its humidity, the very symbol of the *formless*, of the unverifiable, of the nonhierarchized" (Faccini 31). Petroleum is not spittle (nor blood, nor gold, nor shit, nor ejaculate), though these are all versions of the persistent metaphoricity of oil that we will encounter along the way. The commodity's transitive properties (it becomes everything from fuel to plastic to money) and its ubiquity mean that *oil* is a slippery term, transmuting into "black gold" and even transubstantiating into "the very blood of the nation and its citizens" (Apter 277). But petroleum is *like* spittle—the simile captures the havoc these substances play with classification (Faccini 30).

The "literary misfits" (1) Katharine Burkitt reads in *Literary Form as Postcolonial Critique* also "frustrate any easy mode of classification" (2). Burkitt ascribes the unconventional subversions of form in the prose novels, poetry, and epics she studies to their postcoloniality, building on Franco Moretti to argue that "the social aspect of literature *resides in its form*" (2). Formal interventions in the text can spur associative leaps in terms of form for the theorist, too. Stephanie LeMenager, for instance, proposes that we might read novels like poems: "If what a novel produces is a set of lingering sensations tied to the reader's everyday memory through improvised situational analogies, perhaps it can be called a poem, too" (*Living Oil* 130). She suggests this while discussing Helon Habila's 2010 petronovel, *Oil on Water*, and then illustrates it with an even bigger leap: "For example," she writes, "one summer afternoon I 'remember' the Niger Delta that Habila evokes in the smell of spilt gasoline at my local service station in southern California" (130). LeMenager sets off her "remembering" in quotation marks, but they are not sufficient to allow for the equation of these two settings, radically apart in so many ways. Almost

---

[1] Throughout this book, all formatting (such as italics in quotations) is as appears in original texts. Ditto for unconventional grammar, spelling, and punctuation.

certainly the smell of a California gas station is not the same as the smell of the Niger Delta, no matter how universal an odor petrol has. Literature may produce a spontaneous overflow of powerful feelings, but it does not reduce difference to sameness, or risk to safety.

There is no such thing as a singular oil encounter. Philip Aghoghovwia says as much while setting up his analysis of Ibiwari Ikiriko's poem collection, *Oily Tears of the Delta*, arguing "for the imperative of contextual place-based provenances in the representation of human experiences associated with the Oil Encounter" (32–33). Singularity is also undone by the many forms oil itself takes. The various art objects and texts explored in *Petroforms* meet oil as a crude resource leaching from the ground, a refined liquid carried through pipelines, and as gasoline and kerosene in automobiles and canisters. They meet it in different states of matter—as liquid commodity, as waste burnt off in gas flares, and as late-stage solid plastic products. As Christopher Jones explains in *Routes of Power*, it is the very liquid nature of oil that makes it and its industry so environmentally destructive. Comparing the oil industry to the California gold rush, he writes: "The California gold rush also operated according to a logic of economic growth above all else, but neither gold nor the massive quantities of rock removed to find it flowed in the same way that oil did. The fact that petroleum seeped out of wells, ran into streams, and leaked out of storage tanks made it much harder to control" (120). Jones is writing about the new energy transportation infrastructure of early-twentieth-century America, but the same principle holds true universally. It is not just the "materiality of oil" but also a "human disregard for spills" that combine to sacrifice the environment (Jones 120). Petroforms hold the ambiguity of attending to the spills that we normally disregard.

Joshua Schuster, asking "Where Is the Oil in Modernism?," points out, parenthetically, that "(we usually see oil only when something has gone wrong and it is spilled)" (199). His parentheses, partially occluding our view, are emblematic of his argument that

> oil defies direct representation and symbolic narrative. It is hidden underground, the end product of a still not fully known process; the technical apparatuses used to extract it are beyond the knowledge of laymen; to store it is to not see it; and to use it is to vaporize it or fix it into a new material (we usually see oil only when something has gone wrong and it is spilled). Its effects spread inexorably into the conscious and unconscious, begging to be leered at in forms of spectacle and conspicuous consumption but resisting vision all the same. (199)

The idea that oil resists visibility is pervasive in petrocriticism. Contrarily, I argue that oil re/forms aesthetics such that they must make it visible. The immediacy of having to contend with it is quite different from the distance afforded by not fully knowing it. As opposed to what Schuster says about modernism or Ghosh's lamentation in "Petrofiction" about the lack of the great American oil novel—situations in which oil is underground or in the background—these Nigerian objets d'art foreground it. Henry Ajumeze, also writing about Nigerian petrofiction, agrees that "an important aspect of measuring oil's relation with culture is to acknowledge its physicality rather than its invisibility" ("Petro-drama" 100). At the very end of his chapter in *Oil Fictions*, however, Ajumeze cheats the difference between metaphoricity and reality, equating the "aesthetic abstraction of the biophysical property of the commodity through which materiality is metaphorized" with "its ontological presence in the substratum of polluted landscapes in which its damaging smells and decays are tellingly evident" (112). The oil that spills out of petroforms is not an abstraction. Reified, it resolves the "inquiry about energy's visibility or invisibility" (Yaeger 309) because it insists on being seen.

I am aware of the risks of naming the ubiquity of petroleum. If the petroprefix stamps its presence everywhere on everything, it loses its meaning and becomes gimmick instead of signifier. Moreover, petro- as marker does not necessarily connote petroviolence in the way Michael Watts has centered. For instance, introducing petrocriticism to my students in a "Text and Context" course on Ken Saro-Wiwa, I ironically suggested that they might be wearing "petrosneakers." Rather than scoff at the idea, though, they said they might like to buy those sneakers. They are the children of petromodernity, shaped through and through by petrocapitalism. I am also aware of the political risk of seeming to suggest that the corporate and national interests of Big Oil dictate the terms of its own resistance, and of the rhetorical risk of seeming to anthropomorphize petroleum that comes with talking about what it "wants," but I assert that form as a concept lets us see exactly what the corporation nation capitalizes on: that is, the fundamental formlessness of oil, and thus the need for forms to comprehend it and confine it. These risks speak to the need for the project of *Petroforms*—if different forms get at different aspects of living with and around petroleum, it may be that we can only get at the truth of those experiences by examining a set of samples.

Since Amitav Ghosh coined the term *petrofiction* in 1992, the field has practically gone mainstream. Half a dozen and counting "Petrocultures" conferences have been held under the auspices of the Petrocultures Research Group. Petrocritical syllabi are being taught at the University of Warwick, the University of Florida, and Rutgers. Chalmers University of Technology in Gothenburg, Sweden, has a Petrocultures Reading Course. In the Global North, anyway, petrocriticism is rapidly becoming academically institutionalized.

It's far harder to find evidence of such entrenchment in the Global South, however, partly because of a system of "institutionally prepared syllabi prescribed for teaching in universities" (Egya, "Re"). A professor teaching African Fiction at Ibrahim Badamasi Babangida University in Lapai, Nigeria, might assign Helon Habila's novel *Oil on Water* but would not be able to design a "Niger Delta Petrofiction" course (Ega, "Re"). On the flip side, plenty of industry-related courses exist, such as Petroleum Economics I at the University of Ibadan. The course guide for Advanced Oil & Gas Law I at the National Open University of Nigeria acknowledges that "poor management of the petroleum resources has led to socio-economic, socio-political and complex interaction problems involving the people, economic development and the environment" but hedges that critique with both blame (for "rising oil theft and bunkering") and acclaim (for legislations that "effectively regulate the Nigerian petroleum industry)" (Okukpon 6, 7). These industrial curricula aim to interpellate students as Tanure Ojaide's "helmet-wearing graduates." Sule Egya explains:

> The "helmet-wearing graduates" are people of the community who have gone to school, courtesy of the civilizing mission; who have acquired technological skills and are using the same skills, in the thinking of the poet, to bring environmental disaster upon the biodiversity of the land. They are lured into doing this with "foreign currency"—most oil workers in Nigeria are paid in dollars. With utmost enthusiasm and energy, therefore, they embark on activities that they themselves know will eventually harm the environment. ("Pristine Past" 197–98)

We will see degrees of this interpellation in the stories, plays, and films to come.

Nigerian scholars work both ends of the critical spectrum. Petropositive economic, political, and scientific essays about Nigerian oil abound. Petrocritical cultural, social, and literary scholarship carves out spaces for reading against the grain of oil theft and bunkering or the claim that "an abundance of petroleum resources may in fact be much less of a curse and more of a boom for economic performance" (Obafemi et al. 154). Egya, for instance, reads specifically for "literary militants" across Nigerian poetry and novels, while Senayon Olaoluwa draws our attention to the leveling violence of the "oil-induced Anthropocene" (254, 260). What we see in the scholarship, then, repeats in the aesthetic products it analyzes: multifaceted engagements with oil in which cultures of protest about oil extractivism sit alongside cultures of enthusiastic petromodernity.

To teach and theorize Nigerian texts from a distance, as I do, is to indulge in a degree of privileged earnestness that I fully acknowledge. The very existence of petrocriticism as a field is an index of a kind of intellectual bunkering that rarely spills over into actual protest, change, or action. On the other hand, I study literature and culture because I believe that it matters; that it has material impact. I hope—especially in the face of an increasingly brutal full-frontal attack on the humanities—that opening traditional fields and departments to more interdisciplinary approaches in general and to ecocriticism in particular draws curious minds. I also hope that *Petroforms* encourages a cross-cultural, comparative conversation, and that my local focus on Nigerian examples speaks to our global investment in and entanglement with petroleum and its futures.

Amitav Ghosh's *The Great Derangement* (2016) challenges us to think the unthinkable when it comes to climate change and the Anthropocene. He works from the premise that future readers will look back at today's literature and judge us deranged for not having engaged with our climate crisis. But perhaps we should look outside the "mansion of serious fiction" (61) for that engagement. For Ghosh, sci-fi and cli-fi and other subgenres might be dismissed because they are relegated to the "outhouses" (24) of cultural production, but the literature, film, and art I read here indicate that hybrid forms and generic frictions are generative provocations. *Petroforms* is thus a response to Ghosh, but it also picks up on a number of his theoretical threads such as how paying attention to the nonhuman makes us "rethink conceptions of history and agency" (119).

The Anthropocene's use-value as a label and a concept is the subject of much debate, and scholars like Sophie Sapp Moore et al. ("Plantation Legacies," 2019) and Kathryn Yusoff (*A Billion Black Anthropocenes or None*, 2018) agree that its universalizing tendency masks a history of racial oppression. Yusoff writes, "Approaching race as a geologic proposition (or *geologies of race*) is a way, then, to open up the imbrication of inhuman materials and relations of extraction that go beyond a place-based configuration of environmental racism as a spatial organization of exposure to environmental harm." Like Yusoff, Achille Mbembe in *Necropolitics* (2019) wants to expose how the human-as-object is the foundation of not only our economic and political practices, but of our intellectual practices as well. Together, these theorists help us understand the various petroforms read here as interventions in the assumptions and behaviors of the Anthropocene, and also in how it gets defined, and by whom.

Petroforms builds on groundbreaking thinking in these areas and in their subfield, petrocriticism. Amitav Ghosh, Imre Szeman, Michael Watts, Peter Hitchcock, Stephanie LeMenager, and Jennifer Wenzel are touchstone references. Like Wenzel's *The Disposition of Nature: Environmental Crisis and World Literature* this book is also interested in asking "how do literature and other cultural forms shape how we imagine the planet?" (1). And, like Hannah Appel's *The Licit Life of Capitalism: US Oil in Equatorial Guinea*, this is (to quote her publisher) "both an account of a specific capitalist project . . . and a sweeping theorization of more general forms and processes that facilitate diverse capitalist projects around the world."

In Ghosh's "Petrofiction" essay, written in 1992, he arrives at the conclusion that "the truth is that we do not yet possess the form that can give the Oil Encounter a literary expression" (31). The reason, he argues, is that existing literary forms themselves cannot accommodate oil (30). Other scholars have similarly queried whether there are aesthetic vehicles that can adequately capture our ongoing catastrophes, whether they be climate change or the energy crisis (broadly construed, from the instigating insult of extraction to grid failures and beyond). For instance, Richard Kerridge's "Ecocritical Approaches to Literary Form and Genre," in *The Oxford Handbook of Ecocriticism*, ultimately asks, "Where are the rigorously realist novels, with present-day settings, dealing with people's emotional responses to the threat of climate change?" (373). Petroforms respond to that need, often in ways that read as *too* realist for their critics, and they rise to meet the challenge Ken Saro-Wiwa posed in his "detention diary," *A Month and a Day*: "Literature in a critical situation such as Nigeria's cannot be divorced from politics. Indeed, literature must serve society by steeping itself in politics, by intervention, and writers must not merely write to amuse or to take a bemused, critical look at society. They must play an interventionist role" (81). Parsing where aesthetics stop and politics start is an irrelevant exercise when the aesthetics *are* the politics. The forms that carry the weight of petromodernity are so steeped in oil's realities that they overflow.

In her book about energy futures in the Orkneys, *Energy at the End of the World*, Laura Watts builds on Isabelle Stengers's call for composition, not just critique and deconstruction. Watts writes, "I must write a world, not just take the existing one apart" (15). Like Stengers and Watts, I am interested in the constructive constructs of form and reformation, and in my own contribution as world-making, not world-breaking. I believe my close reading methodology makes this possible by opening texts up and inviting us in for interpretation, reshaping our understanding instead of undoing it. To that end, I hope that different readers will engage with this book differently—some may read cover to cover and others will take discrete chapters for specific purposes—to read, to teach, or to make more of. The following is how the chapters unfold.

Chapter 1, "Petrofiction," reads a set of Nigerian short stories by Sefi Atta, Nnedi Okorafor, and others for the ways in which they sabotage Big Oil's narrative and the short story form. Necessarily crude in terms of both form and content, they operate nimbly and tactically, in opposition to the muscular strategies arrayed against them in master narratives that accuse local Nigerians living alongside the sites of oil production of sabotaging or stealing their own resource.[2]

In the course of writing about how Nigerian short stories sabotage Big Oil's narrative, I discovered that romance is the most widely read genre in Nigeria, and that 91 percent of romance book buyers in general are female. My hypothesis was that, given the total saturation of everyday Nigerian life with oil politics, those tensions and debates must inevitably arise in the fiction. As it turns out, while the short stories explicitly call out the dangerous, exploitative nature of the oil industry, in the romance fiction by and for women, the intersections between gender, oil, and the text appear to be far more taken for granted. Romance works within a form to deliberately showcase the pleasures of the text, including the private female pleasure of reading alone, the promise of love and marriage, and the staging of erotic fantasies, which time and again feature elements of petromodernity, from sex scenes in the backseats of cars to flirtations on the side of the road. Chapter 2, "Petrofeminism," lets us see the romance novel as a petroform insofar as it conjures a form of intimacy enabled by petroculture.[3]

Visual art that comments on petroleum practices in Nigeria includes works by Sokari Douglas Camp, who incorporates the industry's materials and detritus into her sculptures. Chapter 3, "Petroart," uses her work as a case study to ask, what is (anti)canonical? What is translated via form, from painting to three-dimensional object? How do the industrial and the natural relate? Her work sets out to decolonize the classical and to find a new, concrete form for marking where oil touches the human. Her public art projects include *Battle Bus: Living Memorial for Ken Saro-Wiwa* (2006), a full-scale bus replica that was impounded by authorities at the port in Lagos on grounds of its "political value" when she tried to import it as part of the twentieth anniversary of his murder. With *Battle Bus*, Douglas Camp transforms a product

---

[2] This chapter first appeared in *Postcolonial Text* 11.1 (2016).

[3] Versions of this chapter were included in Balkan and Nandi's *Oil Fictions* collection and presented at the "Pleasure, Arts and the Human in Africa" panel at the 2020 MLA in Seattle.

of our petroculture into a critique of it, literally driving her point home. The coda to this chapter contrasts still, sculptural forms to the movement of a modern masquerade.

In the "Buzz" section of the website promoting the 2009 documentary film *Sweet Crude*, Ralph Nader is quoted as saying, "Add the petrohorrors in the Niger Delta to the 'price of oil.' There is nothing 'sweet' there, but the oil industry's profits." I take Nader's coinage and ask what a petrohorror film would look like—would there be oil monsters rising up out of the depths as the Oil Blob does in *Creepshow 2*, or do the lived horrors of petroleum extraction in the Niger Delta suffice? Chapter 4, "Petrohorror," suggests that we can read documentary films about oil in Nigeria through the lens of horror. Although the "real" of the documentary would seem to resist horror fictions, in what ways do these films employ those same tropes and to what extent is that either a necessary shock or an indulgent danger? The chapter ends by offering Zina Saro-Wiwa's art video *Niger Delta: A Documentary* (2015) as a way to ask, when horror is drained from the documentary, what is left?[4]

Chapter 5, "Petrocinema," flips the script on Chapter 4 and views popular feature films for their documentary-like qualities. We see that Nollywood films (made in the Hollywood/Bollywood film industry of Nigeria) with oil as their focus cannot help but instruct their audiences. By focusing on two Nollywood feature films, Jeta Amata's *Black November* (2012), and Curtis Graham's *Blood & Oil* (2015), I offer a definition of petrocinema as a substantively different form, films so profoundly imbricated with and influenced by oil that they reconfigure themselves to be didactic about petroleum and its dominion.

In the sixth and last chapter, "Petrodrama," I read Nigerian plays that engage with oil as a matter of personal and political interest, paying special attention to representations of women, gender relations, and female playwrights. The focus on women allows us to see the particularities of gendered engagements with oil, extending that priority from earlier chapters, such as in the discussion of romance novels, and it centers a set of texts that otherwise often get sidelined next to those by better-known, male writers. The chapter starts from Brecht's premise that "petroleum resists the five-act form" (*Brecht on Theatre* 30), looking for the ways in which oil forces dramatic structures into excess or containment. It ends with the provocation that the queerness of oil queers gender relations.

---

[4] Earlier versions of this chapter were presented at the CSA conference in 2021 and at the "Petrocultures: Transformations" conference in Stavanger, Norway, in August 2022.

In the spirit of recuperation, restoration, and reclamation, *Petroforms'* conclusion follows Ken Saro-Wiwa's famous sitcom character, Basi, toward a tentative recovery. This is not, however, a simple or reactionary return to prior forms and habits, but rather a creative recovery that makes newness possible. Writing about Saro-Wiwa, Rob Nixon observes that, "across much of Africa the certainty persists that writing can make things happen" (104). In *Culture and Imperialism*, Edward Said says that "land is recoverable at first only through imagination" (221). *Petroforms* argues that we imagine our way to making things happen, to recovery of what's ours, as much by form as by content. Oil reconfigures our lives, and especially the lives of those, like residents of the Niger Delta, for whom it is omnipresent and inescapable, and in return cultural production reconfigures to accommodate, address, and attack it. Form as container spills over with petroleum products. Sometimes that excess is reabsorbed, sometimes it requires a new container, and sometimes it stays messy.[5]

---

[5] As this book goes to press, news outlets report that the Nigerian government has pardoned Ken Saro-Wiwa and the rest of the Ogoni Nine for the alleged crimes for which they were hanged. As surprising as this announcement is, and as much as it may be one small step on the road to recovery, it does not exonerate them. Nor does it begin to repair the damage that Big Oil has done in the region. The toxic mess it created remains.

CHAPTER 1

PETROFICTION

"The oil companies," she said, "they drill our father's farms and they don't give we, their children, jobs. We eat okra, cassava, grown in other parts of the country. We use their yam, plantain and palm oil to cook our *onunu*. There are no fish in our rivers, no bush rats left in our forest. We don't use natural gas in our homes and yet we have gas flares in our backyards. We can't find kerosene to buy and we have pipelines full of the products running through our land. Some of us don't have electricity. Some of us don't even have candles to burn. Are you listening, women?"

—"A Union on Independence Day," Sefi Atta

Textual Saboteurs

In Sefi Atta's 2003 story "A Union on Independence Day," Madam Queen tries to mobilize villagers in the narrator's hometown to stand up to Summit Oil by deploying language that is striking in its fierce bluntness. New Nigerian short stories, more easily available to readers than ever before, call into question the dominant discourse around oil production nationally and internationally as they depict the complex, lived reality of a resource-rich country riven by corruption, greed, and poverty. The struggle over oil resources and rights regularly involves acts of sabotage. In these stories, these acts range from individual villagers siphoning crude oil from pipelines for fuel to organized, militant attacks on petroleum operations, or, in the case of "Independence Day," to women occupying the Summit Oil terminal in protest (a storyline borrowed from real-life demonstrations and protests that made international headlines in 2002). Because of their content, their form, and their distribution, the stories themselves can be considered small acts of sabotage. They are textual *saboteurs*, in both senses of the etymology of that word: to perform or execute badly (as in, "he murdered that concerto") and to destroy willfully, as with tools or machinery ("saboteur"). The stories' formal qualities—or lack thereof—reflect the circumstances of their production as a tertiary byproduct of sabotage. Like the material byproduct, the crude oil to be distributed as part of an informal economy, these stories are unrefined. This textual output, like that commodity form, is more quickly available, less constrained by production, less costly for being raw or unprocessed, and more willfully destructive of disciplinary institutions—corporate, national, and literary.

What we see at work in these stories are truly tactics, in the way Michel de Certeau intended. Borrowing from military parlance to reformulate the relations between power and resistance, de Certeau assigns the term *strategy* to large-scale state, corporate, and other systems, and he uses *tactics* to describe the small-scale subversions that challenge those strategies in his 1988 volume *The Practice of Everyday Life*. Whereas strategy has the advantage of being massive and muscular, the tactic has its small size and flexibility going for it. The very nature of the short story allows it to act tactically in the face of corporate strategy, its content undermining the master narrative that has been carefully crafted by a corporate-state collusion, and its compact form allowing for fairly easy distribution and access, especially for those stories disseminated digitally. Contemporary Nigerian short stories move tactically, thematically and structurally, and are coincident to real-life, everyday tactics used by ordinary Nigerians.

Big Oil regularly accuses residents of the oil-rich Niger Delta region of acts of sabotage, including illegal *bunkering*, or siphoning oil for use or sale, and vandalizing pipelines and facilities. The language multinational corporations use to describe this behavior sets up a narrative of criminality. They talk about the "theft" of oil, of a "black market" (implicitly linking illegality, race, and the color of oil), and both the companies and the press frame the locals in pejorative terms. Shell's website alleges, "Criminal activities including sabotage, oil theft and illegal refining are causing huge environmental damage in the Niger Delta. From 2008 to 2012, these activities accounted for around 76% of the oil that escaped from SPDC [Shell Petroleum Development Company] facilities" ("Shell in Nigeria: Environmental Performance"). Atta's Mama Queen indicts this habit of calling young men "thugs," and reminds us who the true culprits are:

> Young men are kidnapping expatriate employees and demanding ransoms. They are locked up. We call them thugs. Young girls are turning to prostitution to service expatriate employees. They are locked up, too. We shun

them. We say they bring AIDS. Meanwhile, the oil companies spill oil on our land, leak oil into our rivers. They won't clean up their mess. All they do is pay small fines, if they pay at all.

The "meanwhile" in the middle of this diatribe sutures together independent clauses that might not otherwise be connected—things said and done to the community versus things oil companies do. It reveals a consequential, causal relationship between these ideas and seems carefully chosen, as a word about time, to comment on the temporal uncoupling of these experiences in the normative narrative, in which they appear to occupy concurrent but separate time zones. In reality, though, negligence and neglect on the part of the companies has so drastically impacted the area that its environmental degradation has been called ecocide (Johnston). The stories act as a corrective to this narrative (one that commands enormous authority since it is backed by enormous assets). They reframe the actions of impoverished Nigerian citizens as tactical maneuvers by people wielding the only weapon they have—using the tools of exploitation against their exploiters. They expose the absurdity of the multinationals blaming the people for their own poor behavior and the irony of a mindset that frames those people as criminals "stealing" back their own oil.

In 1992, Amitav Ghosh asked why there isn't a Great American Oil Novel (*The New Republic*), coining the term *petrofiction* and launching a new scholarly industry. In his 2012 article, "Oil and World Literature," Graeme Macdonald answers Ghosh's query with the following:

Questions of oil's visibility and configuration in national literary histories, however, needs [sic] to be reconceptualized on at least two fronts: geographic and generic. What constitutes an American (or indeed a British, Nigerian, Iranian, Trinidadian, Russian, etc.) oil text in an age where the circuitry of literature grows increasingly international, and where many arguments have been made in academic circles to pressurize any national literary outlook as limited or, worse, solipsistic? Following this: what specifically constitutes oil literature? Must a work explicitly concern itself with features immediate to the oil industry? Given that oil and its constituents are so ubiquitous in the material and organization of modern life, is not every modern novel to some extent an oil novel? (7)

Although I understand the desire to highlight the globalizing machinations and mechanisms of the oil industry, to say that every modern novel is an oil novel is to evacuate meaning from literature manifestly about oil. The literature read here, primarily recent Nigerian short stories, is demonstrably historically and geographically specific in its concerns, and it does something generically different and worth taking out of Macdonald's brackets to examine.

Rob Nixon called our attention to the long environmental aftermaths of corporate colonialism in *Slow Violence and the Environmentalism of the Poor* (2011), which has two chapters on oil politics. His examples show the burden of ecological degradation that impacts the health and livelihoods of the poor most directly. When Madam Queen lectures the gathered crowd in "Independence Day," she articulates exactly this effect. The first part of her speech, delivered to an initially skeptical but eventually swayed audience, weighs the ironic contradictions of an impoverished oil economy against each other: The oil companies drill on "our father's farm" but "we their children" have no jobs; "we" live on top of untold wealth but have to import their food; "we" have no energy sources, but pipelines run "throughout our land." The second part of her speech accuses the companies of unethical behavior (as they "spill oil on our land, leak oil into our rivers" but "all they do is pay small fines, if they pay at all") that amounts to murder: "Women, listen to me. I'm telling you this: as we speak, we are dying. We are dying of our air; we are dying of our water. We are dying from oil. We are not benefiting from it." In Nigeria, domestic exploitation clearly connects to the vested interests of the multinational oil companies she points to; Shell has at times generated half the revenue of the government, and in turn the government puts its infrastructure and military at the company's disposal. Who benefits is manifestly obvious. Nixon calls this "the ongoing romance between unanswerable corporations and unspeakable regimes" (105), emphasizing by doing so their mutual desire and dependency. Michael Watts calls it "the slick alliance of state and capital" ("Petro-Violence" 3), as though oil lubricates that relationship in an especially slimy way. If this sounds crude, it's because it really is, in its crass exploitation of people and places. Since the state has become privatized in this way, distinctions between nation and corporation become more difficult to discern. We have long understood that the novel plays a distinct role in nation-building (via Benedict Anderson, Timothy Brennan, and others), and particularly in postcolonial national identity formation. Perhaps this is why there have been almost no Nigerian oil novels[1]—because there is no

---

[1] Some novels, like Saro-Wiwa's *Sozaboy*, make no mention of oil but are implicitly, in every way, about it. *Sozaboy* is set during the Biafran War, the scene of an internal struggle over resource control. Helon Habila's 2010 *Oil on Water* stands alone as a novel explicitly about Nigerian oil.

national identity to represent here,[2] but instead illegitimate, fractured, local-ized affiliations better captured by the similarly scaled short story. Where the nation and the novel are monolithic reflections of global capital strategy, the short story is a finely calibrated tactic that sabotages the notion of national coherence. When Ken Saro-Wiwa, who once remarked that "there is no such country [as Nigeria]. There is only organized brigandage" (qtd Nixon 120), was hanged by the state, he cried out, "What sort of a nation is this?" The short stories answer his question as they explode the pretense of national unity and corporate compliance with portrayals of private gain and personal loss.[3]

Non/Fiction

Uche Peter Umez' 2006 story "Wild Flames," available in multiple online ver-sions, dramatizes tensions between villagers, oil executives, and vigilantes; the story rediscovers the ordinary in its exploration of the contradictions, confu-sion, and violence in everyday life in the Niger Delta. It is also an excellent ex-ample of how these stories are sabotaging fictions in the sense of undermining narrative forms and expectations and reimagining how short stories work; in other words, of seeming to execute the story form badly. It is, frankly, a very confusing story to read, with far too many players and parts, but in that way, it captures its own essence.

In "Independence Day," Mama Queen makes reference to "young men . . . kidnapping expatriate employees and demanding ransoms." Oil executive kid-napping, as a strategy of economic sabotage, is a widespread practice. A rash of foreign hostage taking for ransom began in about 2006, the year "Wild Flames" was published, with hundreds of kidnappings, part of a bigger "growth indus-try" (Campbell). "Wild Flames" puts this ripped-from-the-headlines plot dead

---

[2] This is both a product of a history of colonial acquisition and of modern economic circum-stance. Nigeria is an extreme example of an artificially constructed country, having had its borders drawn at the Berlin Conference of 1884–5 by European powers vying for control of territory and resources (then rubber and people) with total disregard for existing political, linguistic, geo-graphical, and other boundaries. The result was a cobbled-together nation that never cohered. Skip through multiple generations of dispossession, first by colonial rulers and then by post-independence military regimes, to the present-day disposition of corporate-state collusion.

[3] Different models describe this process of corporatization: Whereas Nixon predicts a return to nineteenth-century "concessionary economics" where "the nation-state will become ever more marginal to deals negotiated between local chiefs and transnationals" (119), Jean-Francois Bayat, in *The Criminalization of the State in Africa,* identifies an "economy of plunder" in which "state power is the key condition for the accumulation of wealth" because it appropriates national assets and resources (as explained by Federici 78).

center as a group of young men holds Westoil executives hostage while reciting their demands:

> Armed with cutlasses, this group rounded up most of the staff, and held them as hostages until Mr. Blimp, the base manager, Mr. Laggard, the community relations manager, and a pinstriped-suited Nigerian came out. The boys were invited into the boardroom. There was a table in a corner of the air-conditioned room with trays of food and drink. The boys refused the refreshment outright and blurted out their demands:
> - community relations fee of two hundred thousand naira
> - half a million naira for youth development
> - one hundred thousand naira as environment protection levy.

In moments like these, Umez sabotages fiction as his story not only directly engages with current events and crises, but names Westoil directly, without the pretense of a fictional veneer, and then proceeds to list perfectly plausible stipulations. At this point the story seems like a delivery vehicle for local claims and reparations and begs the question of what the role of fiction is in these politically heightened texts. And yet the short stories are not merely fronts for such demands; Umez is not hiding behind the screen of fiction. Even in a scene like this, practicality is counterweighted by the satire embedded in the characters' names, Mr Blimp and Mr Laggard, which lampoons them as bloated, reactionary fools. "Wild Flames" ends with the narrator's future fantasy: "I sat down on the bank of a river and gazed out over the waters. Then I closed my eyes and saw myself in a canoe gliding towards the horizon." This vision can't be itemized.

"Wild Flames" is a work of realistic fiction, but even in the stories we might typically classify as magical realism or science fiction, the fictional genre is regularly broken with to allow for instructive asides. For instance, in the 2007 story "The Popular Mechanic" (which I'll discuss further later), Nnedi Okorafor interrupts her plot line to tell us this:

> Nigeria was one of the world's top oil producers. Yet and still, as the years progressed, the Nigerian government had grown fat with wealth harvested from oil sales to America. The government, to the great detriment of the country, ate most of the oil profits and could care less about what the process of extracting the oil did to the land and its people. On top of all this, ironically, Nigeria's people often suffered from shortages of fuel. (166–67)

Uwen Akpan's story "Luxurious Hearses" (2008) is a story told during the course of a bus ride, sharing with many of the Nigerian stories a petroleum-fueled vehicle for its central plot device, thereby signaling an inescapable, embedded reliance on the stuff. (For example, we have the ironically titled bus "Progres" in Ken Saro-Wiwa's "Home, Sweet Home" [1986], the car that runs out of petrol in "Baptizing the Gun," and Okorafor's *kabu-kabus* [unregistered taxis, here sometimes magical]). Akpan's story does not drop the fiction quite as abruptly as Okorafor's, but turns didactic nevertheless:

> But these were hard times. Due to decades of oil drilling, the soil was losing its fertility. Rivers no longer had fish, and, worse still, repeated oil fires annihilated hundreds of people each time. Shehu, fearing for his cows, moved away from the oil-rich villages to other parts of the South soon after the wedding. When ancestral worshippers began asking people to bring animals to sacrifice to Mami Wata and other deities whose terrains were supposedly desecrated, Father McBride told his faithful to forget the pagans. The problems deepened when little children began to develop respiratory diseases, and strange rashes attacked their bodies, and the natives started running away to the big cities. ("Luxurious Hearses" 213)

A break like this with traditional narrative might signify a number of things: an urgency to convey the real circumstances of this fiction; a desire to do so in an unadorned fashion; and an active disregard for form, style, and genre—that is, shoddy story-telling by design. Interestingly, Nixon identifies a similar aesthetic—even calling it tactical—in Saro-Wiwa's prison diary, *A Month and A Day*, explaining that "the book's disorderly syncretism is partly circumstantial: most of it was spliced together under the stresses of Saro-Wiwa's confinement in a Port Harcourt prison. Yet one senses in the irreverent, breathless bricolage something tactical as well" (123). Normally the diary would be categorized as autobiography, memoir, or general nonfiction, but Nixon recognizes in it the same quality we have observed so far in works we would automatically call fiction—a slapdash suturing whose disorderliness seems political.

Moments like these appear in practically every story discussed here, and they bear strong resemblances to one another. Their presence might gesture toward an international reading audience in need of edification, but they might equally do work for domestic readership, either acknowledging a shared experience, or, perhaps more likely in the piecemeal state that is Nigeria, illuminating (like the repeated motif of gas flares that light up the night sky) for other Nigerians the situation in their own country. A fidelity to reality infiltrates all aspects of the stories, even those that sound fanciful, such as this moment in verse from "Wild Flames": "There was a time he spoke about the future of our village, as though he were reciting an elegy:

> Westoil will impoverish us/ our condition will be worse/ than Oloibiri/ we shall not live as slaves in/ our motherland/ save we act like activists . . . !'

I'd never been to Oloibiri before, but I had heard of it anyway. In my imagination it was a land of fumes and vultures and corpses" (Umez). This is poor poetry, but that may be because the gap between imagination and reality has vanished to the point of imperceptibility—Oloibiri, the place of origin for Nigeria's oil operations in 1956, then abandoned when supplies dried up in the 1970s, *is* a land of fumes, vultures, and corpses. Manipulations of reality within fiction might be a kind of "instrumental aesthetics" (Nixon 109), but they might also be an indication that the very categories non/fiction fail here. This is not about reducing Nigerian literature—or postcolonial writing in general—to the real, the authentic, or the veridical, but rather it is about a realism that captures material, lived, felt, smelt, *sensed* experiences like this one from Atta's story:

> In my hometown we had rainbow colored water. It tasted of the oil that leaked into our well. Bathing water we fetched from a creek. This smelled of dead crayfish. Our rivers were also dead. When rain fell, it rusted rooftops, and shriveled the plants and farm crops. People who drank rainwater swore that it bored permanent holes in their stomachs. Our roads had potholes as big as cauldrons because of the rain. Only in the villages on the outskirts of town did we have one smooth road. The road ran straight from a flow station to Summit Oil's terminal. The villages had perpetual daylight once the gas flaring started. The flare was where cassava farms used to be. Summit Oil bulldozed those farms and ran pipelines through them. The land was now sinking. The gas flare was as tall as a giant orange torch in the sky, as loud as a hundred incinerators. It sprayed soot over coconut trees. From the centre of town we could smell burning mixed with petrol. People complained that their throats were as dry as if they swallowed swamp mahogany bark. ("Independence Day")

Satire and Sci Fi

"The real is a slippery thing" (24), writes Fred Botting, and the real of oil seems especially slippery; more than one critic has noted that that which seems exaggerated or ironic in petrofiction about the workings of the industry is often actually true. Amitav Ghosh and Peter Hitchcock agree that Abdul Rahman Munif's satire in *Cities of Salt* (1984) falls flat partly because the reality of the situation "pre-empts the very possibility of satire" (Ghosh, "Petrofiction" 9). For Fredric Jameson, "parody finds itself without a vocation" in postmodernism where there is no real to copy; instead, it has been replaced by pastiche, which is the "neutral practice of such mimicry, without any of parody's ulterior motives, amputated of the satiric impulse, devoid of laughter" (17). Regardless of whether satire fails because of too much reality or too little, we must contend with both a long local tradition of satire that seems alive and well in these Nigerian stories and a long global tradition of commodity satire in general and oil satire in particular. For the former, Niyi Akingbe's essay "Speaking denunciation" (2014) usefully walks us through precolonial, colonial, and post-independence stages in Nigerian satire, concentrating on poetry like that of Obiora Udechukwu, who writes in his poem, "The Land," "They stole the people's money/ Lent it back to them/ At 1000% interest/ And they talk of philanthropy" (*What the Madman Said* 54). For the latter, we might reach back to Jonathan Swift in the 1700s and the Corn Law rhymers of the 1800s for clear precedents of commodity satire. If oil as a commodity lends itself to satire, we see that in Upton Sinclair's *Oil!* (1926–27), Munif's book, Edward Abbey's *The Monkey Wrench Gang* (1975), and in the antics of The Yes Men, activists who have posed as oil executives in order to perpetrate hoaxes in which they apologize for Big Oil's behavior: "We are sorry," they announce, "We are sorry for the oil and gas spills that have made your rivers toxic. We are sorry for the gas flares that stink up your villages. We are sorry for the fact that you cannot eat your fish. That you cannot grow on your land, and that you cannot drink your water" ("Shell: We are Sorry," 2:08). In the short stories themselves, we have caricatures of oil executives in "Wild Flames," a snide remark in "The Popular Mechanic" that the disabled father is "*very* lucky" not to have been more gravely injured, and Ken Saro-Wiwa's dark satire "Africa Kills Her Sun," which exposes corruption and ironically anticipated his own death. Satire, itself a mode of slippage between what's said and meant, between the implicit and the explicit, sets out to sabotage the status quo and critique our complacencies. Because of this, it is the perfect monkey wrench in the works of Big Oil's narrative, and while it may well be true that satire sometimes fails, leaving us at an "aesthetic impasse" (Hitchcock, "American Imaginary" 90), it seems like a fitting mode in these circumstances.

Like satire and its "militant ironies" (K. Njogu qtd Akingbe 48), science fiction gets wielded as a weapon of textual sabotage. And a similar critique is launched against them both: Just as satire fails because the truth is already absurd, or so the argument goes, so does science fiction fail for an African audience for whom lived reality is already too much like sci-fi. The real of satire and the real of science fiction get in the way. However, what sci-fi does in these stories is not simply represent a dystopian reality; it does something much closer to what Jennifer Wenzel identifies as "petro-magic-realism" in Ben Okri's work: "If petro-magic offers the illusion of wealth without work, Okri's petro-magic-realism paradoxically pierces such illusions, grounding its vision in a recognisably devastated, if also recognisably fantastic, landscape" ("Petro-Magic-Realism," 457). The difference lies in the speculative nature of science fiction. Nnedi Okorafor's fiction, for instance, suggests a latent future potentiality, turning two kinds of speculatory practice—that of surveillance and control, and that of economic speculation—on their heads.

Okorafor does not imagine a post-oil future like James Kunstler's in *World Made by Hand* (2008), one he is positioned to imagine because he is not writing from a place of total saturation. Nor does she give us a space of exemption as in Helon Habila's *Oil on Water*, where an island is set aside from the mainstream world of oil. Rather, hers is a resource-controlled future, where the pipeline people control their own oil and "The Popular Mechanic" is popular because he is a populist, "a citizen before his people." It is he who first opens the pipeline with his superhuman prosthetic and then pinches it closed so as to protect the gathered crowd from "possible death by incineration," but not before he cautions them: "Go home now. . . . Use and sell what you have taken for good things. Keep your mouths shut and only tell stories of the Igbo Robin Hood Pirate Cowboy Man who took what was owed to him and shared the wealth."

The mechanic loses an arm in a pipeline explosion and, acting as a medical guinea pig, has it replaced with an experimental American cybernetic prosthetic, which he uses to siphon off more fuel for the community. He becomes a cyborg, but the text tells us that he and his fellow villagers were transformed by the technology of oil even before the accident: "It was madness. It's still madness. Look what that damn government has turned us into. Robot zombies scrambling for a sip of fuel!" The mechanic is *mechanized*; in the language of the *OED*, he is "of the nature of the machine" and "worked by the machine." The *OED* also tells us that the mechanic, belonging to the lower classes and because of his manual labor (note the irony of the manual laborer who loses his hand), is vulgar or coarse. The story makes him out to be crude not just adjectivally but substantively. Thrice he equates oil and blood. The first two times, this takes the form of comparison via simile ("'See the color?' . . . 'Looks like blood! Ha! It is diesel fuel.'" and "It was a dark pink, almost like watery blood.") The third time crude *becomes* blood. The mechanic speaks to his daughter:

"You and I," her father said. "We both like to work with our hands. That's why I'm a mechanic and you want to be a surgeon and when you are on break, you climb trees and tap palm wine." He paused. "When someone does something to you and you feel that hot fury, you will react with your hands too. Those Americans were lucky that I chose to spill their pink blood instead of their red blood. Crush necks of steel instead of flesh. Those goddamn Americans. Like vampires, even in the Nigerian sun."

The intimacy of lived relations with oil in the Niger Delta means that the mechanic's "dark brown human flesh" turns into "shiny gold metal," until he realizes that he is still in thrall to the "pink volatile liquid." This substantiates Andrew Apter's theory in *The Pan-African Nation: Oil and the Spectacle of Culture in Nigeria*—that oil figures as blood circulating through the national body—but defigures or disfigures the trope to make it literal again.

In stark contrast to the hazards to which petromodernity exposes the mechanic is the romance of risk conjured by those who benefit from it from afar. In an attempt to comfort stakeholders (because "ONLY A CONCERTED RESPONSE BY ALL STAKEHOLDERS, INCLUDING GOVERNMENT, COMMUNITIES AND CIVIL SOCIETY CAN END THE MENACE OF CRUDE OIL THEFT"), in a digital handout published in 2014, Shell informs its readers that

> SPDC's entire area of operations is covered by pipeline and asset surveillance contracts to ensure that spills are discovered and responded to as quickly as possible. These surveillance activities primarily employ members of the communities the pipelines traverse. There are also daily over-flights of the pipeline network to detect new theft points. ("Shell in Nigeria: Oil Theft")

This particular pdf file was once housed on the Nigeria page ("Nigeria: Potential, Growth, and Challenges") of the environment and society section on the Shell Global site, not to be confused with the Shell Nigeria website. Also not to be confused are "Royal Dutch Shell plc. and the companies in which it directly or indirectly owns investments" since these "are separate and distinct entities. But in this publication, the collective expression 'Shell' may be used for convenience where reference is made in general to these companies." In other words, although "Shell" is widely understood by those whose lives it directly impacts to mean one thing, the name itself is part of a dissembling shell game of quickly moving parts. The website mirrors this discursive sleight-of-hand by constantly moving material around, adding and removing images, and burying information. The strange forms oil produces are not just literary, filmic, or artistic, but also corporate.

In "Spider the Artist" (2008), aside from the fact that robot spiders have yet to be invented, nothing distinguishes the subject matter of Okorafor's narrative from the reality on the ground. In both scenarios, pipeline surveillance is made possible by technological detection. In this story, Okorafor imagines a class of Zombie robots "made to combat pipeline bunkering and terrorism." Says the narrator, who alone among humans forms a bond with them, "It makes me laugh. The government and the oil people destroyed our land and dug up our oil, then they created robots to keep us from taking it back" (104). In the story, Okorafor's robots share the name Zombie with the "'kill-and-go' soldiers" (104)—that is, MOPOL, or the Nigerian Mobile Police, a paramilitary state police force partly dedicated to the protection of oil company assets. In other words, the comprehensive disciplinary techniques of the multinational corporations look very like the doubled, combined forces of the Zombies. In a sudden reversal, however, the Zombie robots "go rogue, shrugging off their man-made jobs to live in the delta swamps" (112). Like the mechanic, they redefine their relation to the machine and as machines, escaping the techno-corporate-industrial complex. In both "Spider the Artist" and in "The Popular Mechanic," Delta residents turn the conditions of production to their own uses and advantages even in circumstances of capitalist-induced crisis and environmental degradation. "We are Pipeline People" (102), says the narrator in "Spider the Artist."

Bunkered

In her 2006 essay "Petro-Magic-Realism: Toward a Political Ecology of Nigerian Literature," Wenzel argues for a connection "between Nigerian literary production and other commodity exports" (449). She focuses on "the troubled state of Nigerian publishing and its fraught relationship to presses and readerships abroad" (455) with a focus on literary prize patronage but if literature can also be considered a commodity, per Wenzel, then Nigerians are interested in controlling that resource as well. UNESCO's 2014 report, "Reading in the Mobile Era," surveyed Worldreader Mobile[4] users in seven developing countries in an attempt to determine who reads what on their devices; they already knew that "hundreds of thousands of people . . . are reading on mobile devices" (West

---

[4] Worldreader was a free app provided by a nonprofit of the same name that is seeking to eradicate illiteracy, since replaced by the BookSmart application. It is worth noting that Nokia, the multinational communications and information technology company, was a partner in the study.

and Chew 9). This, in a marketplace where growth feels like the stuff of science fiction: In the first thirteen years of our millennium, internet usage in Africa grew by 3,606 percent (Matthews). Since then, the continent has added almost another five hundred million users, a third of whom live in Nigeria and Egypt alone (Galal). Most of that use is via smartphone, with over 84 percent of internet traffic in Nigeria generated by mobile devices (Sasu). This is a virtual revolution in terms of access to web content, and bypassing traditional means of publication has significant potential repercussions for literacy and reading rates in Nigeria, which has been called a "bookless country" ("Nigeria Is . . ."),[5] and where illiteracy tops 40 percent (West and Chew 14).

Of particular interest in the study is the fact that readers actively searched for short stories on their mobile devices (West and Chew 52), perhaps landing on *Eclectica Magazine*'s site to find Atta's "A Union on Independence Day" (which can also be found under the name "Independence Day" on *Ìrìnkèrindò: A Journal of African Migration*), or on international science fiction website *InterNova* to find Okorafor's "The Popular Mechanic," or accessing her young adult story "The Girl with the Magic Hands," which shows up at #4 on the list of Top 10 Books Read by Worldreader Mobile users via the app (West and Chew 55).[6] Although readers seek out romance, educational, and religious material more actively, it is worth considering what the short story does that other genres cannot do. In 1842, Edgar Allen Poe, father of the short story, extolled its virtues as the genre that achieves the greatest "unity of effect," which is "preserved in productions whose perusal" can be completed in one sitting ("Twice-Told Tales"). Readers in the UNESCO study, on their phones between ten minutes (for men) and nineteen minutes (for women) at a time, can just about fit a short story into a sitting or two, perhaps during a *kabu-kabu* commute. The short story format accommodates the demands of labor and the allowances for leisure in ways that longer forms do not, and the medium, the cellphone, allows for immediate and casual absorption of information.

If the defining qualities of the commodity form (which we see in commodities ranging from crude oil to the mobile device) include portability, transferability, and exchangeability, then these attributes converge in the short story form, so much so that the story might look like a capitalist artifact. On the

---

[5] Kenyan publisher Henry Chakava has argued that "bookless" societies in the Global North are post-book societies, where the same term in the Global South means pre-book societies (Matthews).

[6] In the years since the report was published, both of Okorafor's stories have disappeared from these sites.

other hand, the story form is in no way commensurate with the commodity form, not least because it is neither uniformly the same nor uniformly divisible. In addition, unlike a commodity, the short story can be quickly distributed and disseminated. Thus it might have the power to generate, through rapid circulation enabled by new technology, alternative solidarities. That said, in a potentially contradictory fashion, these particular short stories—because of their crude nonconformities and lack of coherence—rather than being other to the corporate-state, might actually fit the logic and sensibilities of late capitalism, which prefers no organized counternarrative.

We turn to theories like Peter Hitchcock's in *The Long Space: Transnationalism and Postcolonial Form* (2010) to understand extended works of postcolonial fiction as part of a protracted engagement with decolonization, and we turn to short stories for examples of quick, tactical engagements. These may not be on the scale of subversive collective agency, but they might, nevertheless, offer glimpses of connection above, beyond, or outside the state. The internet is a burgeoning site of access for Nigerians to stories about themselves; 14 percent of searches in the UNESCO study were for reading material "from my country" (West and Chew 57). Since the extent to which Nigeria exists as a country has been called into question, and since *country* is UNESCO's choice of words, we can interpret it very loosely to mean less nation-state than land or locale. Elsewhere, I have argued that the postcolonial nation recognizes itself partly through domestic tourism, as locals gain the freedom, mobility, and leisure to tour their own country.[7] We might understand the phenomenon described in "Reading in the Mobile Era" as domestic literary tourism, working in much the same way by letting local readers see their land for themselves. Less liberatory though, is the possibility that part of what sells in tourism and in literature is violent, macabre, and graphic content—the stuff of thanotourism,[8] risk tourism, and war tourism. There is a fundamental difference between a local Nigerian reading of "rainwater . . . that bored permanent holes in their stomachs" whose experience of violence is affirmed by it and the first-world reader who gawks at the description. Oil that is illegally siphoned by militants and vigilantes often ends up being reabsorbed into the corporate-national structure of exploitation—does something akin to that happen to these stories? When Oprah picks Uwen Akpan's story collection, *Say You're One of Them* (2008), for her book club, does that mainstreaming negate whatever subversive, tactical advantage the stories might have had?

---

[7] In my first, book, *Postcolonial Nations, Islands, and Tourism.*

[8] *Thanotourism*, visiting places of death or tragedy, is usually associated with historical sites and memorials. It is interesting to think of it here for places that are actively being traumatized.

The everyday practice of siphoning or bunkering fuel denotes a different kind of reappropriation. Bunkering as a term has suggestive associations: traditionally, on board ship, you fill bunkers with fuel for your own consumption (here this becomes taking fuel for yourself), and it can also mean being placed in a position difficult to extricate oneself from ("bunker"). Read one way, Nigerian locals are clearly bunkered by their circumstances, but read in reverse, the oil companies find themselves at the mercy of bunkering, which we see in their high anxiety about loss of revenue to the practice, manifested in ever-escalating accusations. The Shell "Oil Theft" pdf contends that[9]

> Crude oil theft, sabotage and illegal refining are the main source of pollution in the Niger Delta today. In 2013 the Nigerian government estimated crude oil theft and associated deferred production at over 300,000 barrels of oil per day (bopd). Intentional third-party interference with pipelines and other infrastructure was responsible for around 75% of all oil spill incidents and 92% of all oil volume spilled from facilities operated by the Shell Petroleum Development Company (SPDC) over the last five years (2009–2013). Much greater volumes of oil are discharged into the environment away from SPDC facilities through illegal refining and transportation of stolen crude oil. In 2013 the number of spills from SPDC operations caused by sabotage and theft increased to 157, compared to 137 in 2012, whilst production losses due to crude oil theft, sabotage and related temporary shut-downs increased by around 75%. On average around 32,000 bopd were stolen from SPDC pipelines and other facilities, whilst the joint venture lost production of around 174,000 bopd due to shutdowns related to theft and other third-party interference. This equates to several billion dollars in revenue losses for the Nigerian government and the joint venture.

A very keen sense in this of leaking value shows up in both the figures and the diction ("theft" alone is used five times here). It clearly conveys the perception (however insincere) that "third-party interference" poses a clear and present danger to the SPDC. The final sentence of "Spider the Artist" reads, "You should also pray that these Zombies don't build themselves some fins

---

[9] We cannot trust in any of these numbers presented as fact despite the nifty graphic provided alongside them (they are as fictive as anything Okorafor writes): Shell has been repeatedly made to recant its claims in recent years as Amnesty International and other groups contest them. Shell is motivated to assign blame for spills elsewhere because it bears no legal responsibility for damage due to sabotage.

and travel across the ocean" (115). What sounds like a figurative threat in fact reflects the scope of local sabotage; as the *Wall Street Journal* puts it, bunkering does have "far reaching consequences for global oil supplies" (Faucon). We recall that the word *sabotage* traces its roots back to the shop floor, where workers, in their heavy *sabots*, maliciously damaged or destroyed machinery during labor disputes. By extension, we can think of the people on the ground, the "pipeline people," in an equivalent labor position, and the authors of these short stories as cultural workers in their own right, positioned at the moment of production to disrupt and delay the flow of commodity and capital.

Although Big Oil's master narrative would have us believe that it is made economically vulnerable by sabotage and personally vulnerable by executive kidnappings, the aesthetic risks these fictions take are more than just analogous to economic speculation or a risky investment; they expose who is really at financial and personal risk (that is, who is most likely to be indebted and injured) in the structures of petrocapitalism.[10] The mechanic

> was burned on his face and his entire right arm was burned to the bone. He was *very* lucky. He'd not been that close to the gasoline pool and the burst pipe, and the bodies of the plump laughing women in front of him had shielded him from the giant fireball that rushed past them all like an unleashed demon. All of those women died. Anya's father was one of fifteen survivors. All together, ninety-nine people were killed in the explosion, including an infant who'd been strapped to her mother's back. (Okorafor, "Popular Mechanic" 3)

The narrator of "Independence Day," trained as a nurse in Nigeria but working as a nanny in the United States, tells us generally about "patients with strange growths, chronic respiratory illnesses, terminal diarrhea, weeping sores, inexplicable bleeding. We had too many miscarriages in our town, stillbirths, babies dying in vitro, women dying in labor. People blamed the gas flare" and more specifically about "the story of one little rascal nicknamed

---

[10] There are political risks as well, though while it's true that "in countries like Nigeria where official brutality and paranoia feed off each other, unofficial writing begins to assume the status of latent insult" (Nixon 121), and that, as a result, writers of all stripes have been censored, arrested, and, as in the case of Saro-Wiwa, worse, it's also true that we must stave off the temptation to equate political risk and worthwhile writing. Even a critic like Wenzel, who writes thoughtfully about how Nigerian writing gets assessed and rewarded, seems to suggest a connection like this in her review of *Oil on Water*, when she notes that, "In addition to the Caine Prize, Habila won a Commonwealth Writers Prize for *Waiting for an Angel* in 2003, a feat that he repeated in 2011 with *Oil on Water*—without even getting arrested" ("Behind the Headlines").

Boy-Boy. Boy-Boy wore glasses that belonged to his dead grandfather. He was always with his homemade catapult trying to kill birds. He burned in a gas flare fire. His family held a funeral for him. They had nothing but his ashes to bury. They buried them in a whitewashed wooden casket" (Atta). The Catholic priest who narrates Uwen Akpan's "Baptizing the Gun," a story predicated on a fundamental misreading of a situation as violent because of a culture of violence, has come to Lagos "after an oil fire killed hundreds of my fellow swamp-dwellers in the Niger Delta, after the mass burials, after negotiating with the leaders of the scores of tribes that make up our church to insure that everybody's burial ritual was presented during our week of mourning" (1). This catalogue of fictional hurts matches real damage that will not ever be sufficiently recompensed with financial damages[11] (if such a thing is even possible) and might even be further capitalized on. Plenty of evidence shows how Big Oil and other global players profit off of poverty and pollution—for instance, large-scale bunkering lets companies extract crude in excess of their licensing agreements under cover of sabotage; proceeds from small-scale bunkering go to buying other commodities on the global market like guns and drugs, which in turn further destabilize the region; and general environmental degradation allows for the absorption of spills, decaying infrastructure, and other leavings of resource exploitation (Asuni). The destruction of the region is thus not just a waste byproduct of a capitalist machine but further speculation in it. Okorafor's mechanic embodies this phenomenon when his tragic loss becomes a new investment for "the American scientists" ("Popular Mechanic" 1) who "had since worked out enough of the kinks to safely experiment on their own citizens" (4). These are the material equivalents of global capital's preference for narrative instability.

Necessarily Crude

Critiques of the stories as flawed, imperfect, even failed narratives arise frequently in discussions about them. This might be one strategy for dismissing or diminishing them, and to the extent that it is in reaction to their bald realism,

---

[11] In 2015, Shell agreed to pay a Delta community $80 billion in compensation for two oil spills, but according to one report, they were "upbeat" at the ruling because they believed it would ultimately be resolved in their favor (Oyibo). The scale of the problem—Shell admitted to more than 550 spills in 2015 alone, according to Amnesty International ("Nigeria: Hundreds")—coupled with the protracted legal disputes over each claim, bodes poorly for any hope of real reparations. In that same year, Shell conveniently divested itself of a 30 percent interest in its decaying assets in the region ("Oil Theft"). The 2015 case remains unresolved, although a 2024 United Kingdom Court of Appeals ruling in favor of the Delta communities offered hope that it would go to trial (Bojang).

it does succeed in focusing attention on their particularities rather than on the way in which they comment on macro processes and international politics. An unsigned online review of Akpan's "Baptizing the Gun" disqualifies it as fiction, saying, "If we have not just been told something that happened to Uwem Akpan himself, then we've heard a story told by one of his colleagues. The varnish is pretty thin." In Okorafor's author's notes at the back of her 2013 collection *Kabu-Kabu* (which reprints "The Popular Mechanic" and "Spider the Artist"), she writes of "Icon," "This is a story that is an even mix of the fictional and the real. Some of the words in this story come from individuals in Nigeria's conflicted Niger Delta whom I had no business speaking to." Self-deprecatingly, she also asks her readers to "please be kind with" the first story she ever wrote, and of another she says, "I must admit, looking back, I don't like this story very much." But what these (self)critiques find aesthetically unsophisticated in the writing is also what makes them compelling; they explicitly extract from their circumstances what is *crude*, in every sense of the word. In Akpan's "Baptizing the Gun," the narrator priest suspects a Good Samaritan of being a scam artist and a thief, worrying, "Maybe I shouldn't have accepted the watch from the oil-company executive who attends my church in the first place."

The stories—and petrofiction in general—make us as literary critics struggle with form and genre in a way that puts us on notice. Throughout recent scholarly production in the field, the same struggle is called by different names. Ghosh, writing both as critic and as novelist, confesses, "I can bear witness to its slipperiness, to the ways in which it tends to trip fiction into incoherence" ("Petrofiction" 3), and goes on to say "the truth is that we do not yet possess the form that can give the Oil Encounter a literary expression" (4). Hitchcock, taking up Ghosh's gauntlet (82), argues "it is oil's saturation of the infrastructure of modernity that paradoxically has placed a significant bar on its cultural representation" (81). His essay, "Oil in an American Imaginary," reflects on "the question of an appropriate cultural form" (94) for our petromodernity. Macdonald, in turn, critiques Hitchcock for "viewing the problem as one of pervasive mystification" ("Resources of Fiction" 7) even though he himself writes, "the fantasies of oil culture continue in part because . . . oil *is* fantastic" (17). To some extent, that slippery quality inherent to all aspects of oil—its material being, its extraction and production, its sales and advertising, its taken-for-grantedness—explains the category crisis it produces in cultural representations and their analyses. Across the critical board, fault is found in oil fictions even as those flaws are celebrated. Thus, they have "chaotic surfaces," "many critical detractors who question whether these are novels at all," and a "calculated messiness" (Hitchcock, "American Imaginary" 84, 85, 89).

In this way, they exist on a continuum with a postcolonial vexing of form as habitual practice.

The Nigerian short stories embrace this crisis. Although they are eager to explore oil's metaphorical tendencies (figuring it as blood, for instance), they do not sublimate it. Although they present themselves as fiction (by the way in which they are classified online or gathered together in short story collections), they experiment with bald facts, background histories, and direct accusations. They do not always speak explicitly about oil, but they are always literally about oil. To have an "energy unconscious" (306) in the way that Patricia Yaeger means is a luxury these stories cannot afford—they are utterly and completely energy conscious, which is not to say that they are not simultaneously other things as well: satirical, fictive, and science fictive. "Oil . . . clearly has form," says Macdonald, but it has not been fully realized ("Resources of Fiction" 9). Is he talking about the material stuff that is oil or its textual presence? Oil itself, of course, has substance but clearly has no form—as a viscous liquid, it takes on the shape of its container. If oil as materiality generates specific formal effects, then contained by Nigerian writing, oil's form has been realized as necessarily crude.

CHAPTER 2

———

PETROFEMINISM

———

Don't we all in the end write about love? All literature is about love. When men do it, it's a political comment on human relations. When women do it, it's just a love story. So, although I wanted to do much more than a love story, a part of me wants to push back against the idea that love stories are not important. I wanted to use a love story to talk about other things. But really in the end, it's just a love story.

—Chimamanda Ngozi Adichie to Emma Brockes

What Women Want

When writing *Americanah*, Chimamanda Adichie self-consciously "wanted to use a love story to talk about other things," but love stories are always about other things. For Nigerian love stories—whether mainstream romance novels from the South or the "love literature" of the Islamic North—one of those other things is oil. To expand on Hitchcock's generative law (mentioned in my introduction): oil not only generates but regenerates and is gendered. Here, I read Nigerian romance fiction for what it says about gendered relations to oil, showing that oil is indeed everywhere and making that ubiquity obvious. In what follows, as I move through a series of romantic motifs that emphasize the pleasures of and in the texts, those points of intersection become visible either because the text puts them on display or by virtue of us being attentive to them in ways the text might not anticipate or think noteworthy. This approach, in revealing the petroculture that intimately structures our daily lives and loves, lets us see romance as one refinement of our petro-imaginary and petrofeminism as a necessary development in theories of petrofiction that drastically underrepresent women as consumers, producers, and reproducers of oil and petroculture.

When Janice Radway's now-classic *Reading the Romance* was first published in 1984, it gave theoretical form and expression to a genre and its readership who had been ignored by academia. Radway was interested in "what women want from romance fiction" and how they are able to "circumvent the industry's still inexact understanding" of that (47). In the course of conducting interviews with her informants, Radway came to realize that the "act of romance reading" (18) mattered more to them than the content they were reading; what they wanted was escape from the day to day (a temporal condition characterized by a lack of privacy, sparse free time, and frequent interruptions) (87). Radway describes the books as "eagerly read . . . by women who find quiet moments to read in days devoted almost wholly to the care of others" (87). Her women read for the same compensatory reasons—and to escape similar circumstantial limitations—that the Victorian woman read *Pride and Prejudice* or *Jane Eyre*. *Jane Eyre*, in fact, opens with a scene in which the private female pleasure of reading alone is violently interrupted by a male character. Although we certainly don't want to collapse the experiences of white nineteenth-century Englishwomen, white twentieth-century Midwesterners, and Black for consistency twenty-first-century Nigerians, their reading practices are nevertheless on a continuum, even allowing for radical differences in location, literacy, and leisure.

Gendered relations to oil remain understudied, though important work has been done, such as Heather Turcotte's work on petrosexual politics that argues that gender violence is a necessary precondition for petroviolence. Sheena Wilson looks at how images and concepts of women are "systemically co-opted to serve national and international petro-politics" (177). What I'm calling *petrofeminism*[1] may be more immediately visible in the solidarities and alliances forged through and modeled by explicitly political fictions about oil, such as one of the stories discussed in the previous chapter, Sefi Atta's "A Union on Independence Day," which uses elements like realistic news headlines ("Nigerian Delta Women in Oil Company Stand Off") to deliberate effect. But romance fiction, despite its many retrograde qualities, may also be a site of feminist solidarity, whether through its readers who share an interpretive community, or through its writers, as in the case of the Kano group in northern Nigeria, writing in the shared space of the cooperative and finding self-determination in adult literacy and professional success. When we look for

---

[1] To the best of my knowledge, this term is my coinage. Since I first used it in an early version of this essay, presented at the fifteenth Annual CSA conference at Georgetown University in May 2017, it has started to enter the discourse, although not always with my denotations.

this particular strand of feminism, shaped by, reactive to, and corrective of a petroarchy[2] in the romance fiction, it suggests that romance, love, and pleasure are as much oil relations as are dirt, violence, and degradation. Petrofeminism highlights the constructive potentials of writing, love, and care in the service of various kinds of liberation ranging from the individual (having to do with selfhood and sexual identity) to the epochal (having to do with anthropocenic concerns like mobility and energy independence). Because both petrofeminist theory and fiction are site-specific, we must speak of petrofeminisms, plural. In Nigeria, concerns about the importation and imposition of white Western feminism apply to Adichie and her work,[3] and the African-Islamic feminism of northern Nigeria, as Shirin Edwin calls it, will be different again from feminist expressions in the South. Petrofeminism, while being a retort to globalized corporate self-interest, might also turn out to be a homegrown feminism appearing in homegrown writing.

A search for "Nigeria" in the romance imprint Mills & Boon's website returns two hits, both for books starring Mack Bolan, a heavily serialized character who fights terrorism the world over. In *Insurrection*, Bolan must "smash al Qaeda's hopes of building yet another major African power base" and the blurb for *Conflict Zone* tells readers that "Nigeria is rich in oil, drugs and blood rivals—on both the domestic and international fronts. Mack Bolan's ticket into the chaos is a rescue operation involving the kidnapped daughter of an American petroleum executive." We know that Nigerian women are reading Mills & Boon books[4]—Adichie, in an interview with Zadie Smith, reported that "every girl who grew up in Nsukka when I was growing up read Mills & Boon. I think maybe I read 200"—but they are not finding themselves represented in them. There appear to be no Mills & Boon romance novels set in Nigeria, and even the Mack Bolan series is not strictly romance but action-adventure

---

[2] Petroarchy is a term sometimes used to describe a society governed by oil interests. Here, I am thinking of where that coincides with patriarchy.

[3] See, for example, Damilola Odufuwa's lengthy piece for CNN called "Nigeria: The Women Who Reject Feminism" for how this debate has played out on social media among Nigerian celebrities and politicians.

[4] In second place on the list of Top 10 books read by Worldreader Mobile users in UNESCO's "Reading in the Mobile Era" report is Mills & Boon's *The Price of Royal Duty* read by 18,364 people from seven countries (including Nigeria) in three months. Mills & Boon books are organized thematically on their website; for what it's worth, *The Price of Royal Duty* appears under "sheikh," a series overflowing with oil references in which characters say things like, "'Something always draws me back to the oil fields. A need to keep the rigs safe. And a sense of need to return burning wells to productivity. Duty, passion. I'm not totally sure myself'" (McMahon 119).

published by another division of Harlequin Books. What I read in this essay, therefore, are not Western representations of Nigerian romance, but a spectrum of Nigerian ones, from the self-published *A Heart to Mend* (Myne Whitman,[5] 2009); to *Americanah* (Chimamanda Ngozi Adichie, 2013), a novel that, through acclaim, has risen above the stigma of romance into the stratum of literature; to *Sin Is a Puppy That Follows You Home* (Balaraba Ramat Yakubu, 1990), the first Hausa novel of any sort translated into English.

In *Americanah*, oil fills in background details, a given, constituting part of the market economy in the sections of the book that take place in Nigeria, expressed in the hope that an oil company would rent a block of flats ("'Don't worry . . . God will bring Shell'" [27]) and in dialogue overhead at a party ("'But you know that as we speak, oil is flowing through illegal pipes and they sell it in bottles in Cotonou! Yes! Yes!'" [37]). In her essay on the novel, Katherine Hallemeier remarks on how it marginalizes global economic history in favor of the central love story: "By treating the political economy as a minor plot point in the romance between Ifemelu and Obinze, Adichie's novel belies expectations that African literature ought to do otherwise" (236). The book, while not marketed as a romance, has many of the classic hallmarks of one[6] and has been called one in numerous reviews.[7] Adichie herself has both claimed and rejected that label, saying, "This is in the grand tradition of Mills & Boon but also it's the anti-Mills & Boon" (Smith and Adichie). When Ifemelu and Obinze meet at a dance party, Adichie makes a similar linguistic move, putting the idea of romance under erasure by writing, "Ifemelu thought Mills and Boon romances were silly, she and her friends sometimes enacted their stories." Despite Ifemelu and her friends thinking romance novels are silly, they reenact them anyway. The run-on sentence above strikes me as

---

[5] Nkem Okotcha's pen name.

[6] It is what *Writing a Romance Novel for Dummies* (Wainger 323) calls a "reunion romance," also known as a second-chance romance, and closely follows Shoshanna Ever's "Secret Formula of Most Romance Novels" including Boy meets Girl (Ifemelu meets Obinze); Girl has a secret (she exchanged sexual favors for money); Boy returns, repentant, to declare what they both knew all along (that he loves her); and Girl is now strong enough to turn him down or take him back as an equal partner (*Americanah* ends with Obinze appearing at Ifemelu's door and her inviting him in).

[7] Additionally, critics have pointed out the "(surprisingly spare) critical commentaries" on *Americanah* (Jennifer Leetsch, "Love, Limb-Loosener" 3), with one attributing this directly to its romance form: "The reasons of the lack of academic interest in *Americanah* may lie in the very nature of the novel, the plot being quite linear and 'pop' and based on a romantic love story; this aspect, however, may be considered on the contrary positive, since it makes complex gender and 'racial' issues more accessible and visible to common readers" (Scarsini 11).

purposeful; these things coexist in the same breath without contradicting each other. By making direct, if ironic, reference to the genre, the author flirts with it:

> Ifemelu thought Mills and Boon romances were silly, she and her friends sometimes enacted their stories. Ifemelu or Ranyinudo would play the man and Ginika and Priye would play the woman—the man would grab the woman, the woman would fight weakly, then collapse against him with shrill moans—and they would all burst out laughing. But in the filling-up dance floors of Kayode's party, she was jolted by a small truth in those romances. It was indeed true that because of a male, your stomach could tighten up and refuse to unknot itself, your body's joints could unhinge, your limbs fail to move to music, and all effortless things suddenly become leaden. (69–70)

Both character and author appropriate the romance novel. For the character, romance novels permit safe experimentation with gender identity and budding sexuality; for the author, they allow a self-aware cooptation of the genre's discourse: "'This is really corny but I am so full of you, it's like I'm *breathing* you, you know?' he had said, and she thought that the romance novelists were wrong and it was men, not women, who were the true romantics" (239). Later, we'll see a similar moment in Whitman's *A Heart to Mend*, suggesting self-conscious romance as a postmodern genre, registering itself and reflecting on the act of reading.[8] The postmodern romance nudges the pleasure of the text from the readerly toward the writerly. After dancing, Ifemelu next tells Obinze that what he said sounds like "the kind of thing you read in a book" (70), and Yogita Goyal notes that "questions about reading and reception are themselves staged in *Americanah*, which embeds an ongoing critique about books, how they're read, and what they do or fail to do in the world" (xiii). She goes on to say of Ifemelu: "Each of her romances . . . is mediated by a set of reading protocols" (xiii). With its interspersed blog posts (extended beyond the bounds of the traditional novel form as an actual blog embedded in Adichie's website), the

---

[8] For connections between postmodernism, feminism, and romance, see Diane Elam's *Romancing the Postmodern*, although Elam does not discuss romance novels per se, except by way of comparison to other literature she reads (Umberto Eco, Joseph Conrad, Walter Scott). For Elam, the romance is postmodern because it marks and is marked by excess—"its capacity to appear where least expected" (12)—but this theory only works when one is looking for romance outside of romance novels.

book upends any usual protocols, in keeping with what—and how—women really want to read now.

Being Moved

Perhaps Nigerian women are reading Ifemelu's blog on their phones. UNESCO's 2014 "Reading in the Mobile Era" report, discussed in the previous chapter, describes the ways in which the globally ubiquitous mobile phone is already being used as a reading device and could be further leveraged as a delivery system for books:

> The study shows that mobile reading represents a promising, if still underutilized, pathway to text. It is not hyperbole to suggest that if every person on the planet understood that his or her mobile phone could be transformed—easily and cheaply—into a library brimming with books, access to text would cease to be such a daunting hurdle to literacy. An estimated 6.9 billion mobile subscriptions would provide a direct pipeline to digital books. (West and Chew 17)

A couple of things are worth noting about this excerpt: While "pipeline" may simply be metaphor, it may also not be, especially considering that the report cites the Global System for Mobile Communications (GSMA) here (and frequently throughout), with the GSMA being "a trade body that represents the interests of mobile operators worldwide." It is undeniably in the GSMA's best interests that more phones are used and that more uses are found for phones. I am reminded of Wenzel on "literature as one commodity among others" ("Petro-magic-realism" (450). Not only is the cellphone a commodity, but it is a petroleum product, and a report like this inadvertently (or not) promotes the exploitation of an emerging African marketplace for a number of commodities in conjunction—the book, the mobile phone, the car, and oil.

The "feminist automobility" (161) Lindsey B. Green-Simms theorizes in *Postcolonial Automobility: Car Culture in West Africa* is closely adjacent to my theory of petrofeminism. Green-Simms opens with a discussion of car culture in *Americanah*, noting the heightened relationship between the Nigerian characters and their cars: "Unlike the Nigerian characters, the American characters are never introduced along with the type of car they drive, they do not seem to be particularly consumed with the need to acquire new cars, they do not carry on long conversations in stalled city traffic, and they almost never talk about infrastructural failures" (*Postcolonial Automobility* 3). In many ways, her book details the West African—and specifically Nigerian—romance with the automobile, and, by extension, with petromodernity in general, including the "romance of the national highways—celebrated in conversation, oral narratives, and song" (Mark Auslander qtd Green-Simms, *Postcolonial Automobility* 211). Multiple mobilities—or, at least, the aspiration to them—intersect in the romantic space carved out by the collaborative technologies of mobile phone and automobile. They both hold out the "promise of autonomous, unfettered mobility" (Green-Simms, *Postcolonial Automobility* 3) and a liberation from place, especially places associated with confinement and domesticity.[9]

What Radway calls the "event of reading" (7) is always contextual, and in the case of Nigerian romance readers, oil literally greases the engine of the text's conveyance since Nigerians read while mobile in more than one sense. They read on their mobile phones and they read on the move, while commuting. As I mentioned earlier, Nigeria's mobile phone penetration is staggering, at an estimated 91 percent of the population (Statista), and the country has one of the highest mobile and online readerships in the world (Onwuegbuchi). The UNESCO report has therefore been outpaced statistically but anecdotally it is still germane. It invites us to "Meet Nancy,"

> a mobile reader in Abia State, Nigeria. Nancy is 20 years old and loves to read. Nancy's favourite book is *A Heart to Mend* by Nigerian romance author Myne Whitman. Nancy began reading on Worldreader Mobile in May 2013, and that month she spent 10 hours reading. In June, Nancy read on Worldreader Mobile for over 40 hours. When asked 'Do you think that you read more now that you can read on your mobile?' Nancy replied, 'I do not think that I read more—I know that I read more.' (West and Chew 47)

Nancy testifies to what Emily S. Davis calls "the new intimacies generated within and through electronic media" in *Rethinking the Romance Genre* (162). The dual technologies of the mobile phone and the car conspire to associate reading, romance, and petromodernity in ways that register in the literature both blatantly and latently, and in both event and content.

Wendy Griswold, author of *Bearing Witness: Readers, Writers, and the Novel in Nigeria*, for instance, informs us that the single most common scene in the Nigerian novel is the traffic jam. Indeed, in Whitman's book (freely available

---

[9] We see these come together in a moment to which Green-Simms draws our attention: "Ifemelu's previously unemployed father has a job at a bank and one of the first things he does is to purchase new tires for her mother's car and buy a mobile phone, ensuring the family's mobility" (*Postcolonial Automobility* 2).

to download or read online) much of the connective tissue of the plot involves anticipating traffic, negotiating traffic, or sitting in traffic. Our protagonist Gladys arrives in Lagos on a "rickety yellow bus" that takes an hour to cross the bridge into the city, affording a view from its "smudged windows" of a polluted petroscape that will form the backdrop to the story:

> Men, women and children paddled canoes on the murky black water beneath; oblivious to the stench Gladys could smell from the top of the bridge. The large numbers of wooden houses balanced on stilts above the lagoon had rusty tin roofs wreathed in coils of smoke from open cooking fires. A sawmill further down burnt sawdust and belched up even more clouds of smog. All this mixed with the exhaust plumes from the traffic to create a dense haze that tickled her nostrils. The late January sun outlined the façade of the Lagos Island skyline that towered in the distance. The many high-rise buildings may have been a different color in an earlier life, but all she saw now were structures blackened by decades in the fog. (Whitman 12)

Nigerian traffic has taken on mythic proportions, described in media accounts such as *The Atlantic*'s "World's Worst Traffic Jam" article as an "apocalyptic scene" (Hammer); the language is in keeping with a trend to depict African realities as dystopian sci-fi futures that have already arrived.

It follows that in what is not just a car culture but a culture of "chronic gridlock" (Hammer), the portability of the mobile phone fits. As though by way of illustrating this, in *Americanah*'s frame story, when Ifemulu tells Obinze she is moving back to Nigeria from America (thus becoming an Americanah), he reads her email while stuck in Lagos traffic. As an Americanah, one who "look[s] at things with American eyes" (Adichie 475–76), Ifemelu feels "assaulted" by the city upon her return, including "the yellow buses full of squashed limbs" and the pervasiveness of mobile phones: "When she left home, only the wealthy had cell phones, all the numbers started with 090, and girls wanted to date 090 men. Now, her hair braider had a cell phone, the plantain seller tending a blackened grill had a cell phone" (475–76). Proof that cell phones are everywhere and, more pertinently, that they facilitate romance.

The UNESCO report is enlightening about who reads what on their phones. Clearly, readers want to be emotionally moved while physically moving, since romance tops the list, with phrases relating to romance (e.g., *sex, love, Romeo and Juliet*) among the most popular search terms, and nineteen of the top forty books read being romance novels (West and Chew 52). If genre is a set of rules for the production of meaning (Morley qtd Radway 10), then the rules of romance novels show up across the genre, regardless of whether they are from the Harlequin "fantasy factory" (A. Jones 211) or homegrown Nigerian ones (Ifedigbo).[10] Pamela Regis, in *A Natural History of the Romance Novel*, outlines eight essential elements all romance or courtship novels share. However, although the conventions are persistent, they are not static, so that, for instance, even *Jane Eyre* "demonstrates the flexibility of the form" (Regis 85). Ann Rosalind Jones has studied how Mills & Boon novels absorb and accommodate feminism. Jones surmises that the romance format is particularly rigid when it comes to the conventions through which the hero is constructed: He is still older, richer, wiser in the ways of the world, and more experienced sexually than the heroine. In the Nigerian fiction, these rules hold but the "urban petroscape" (LeMenager, "Aesthetics" 66) modifies them. Scenes that associate the eligible, desirable hero with his car drive the narrative, and while the narrative perspective still privileges the male gaze (A. Jones 214), he now habitually looks at the woman on the side of the road from inside his luxury vehicle.

In *A Heart to Mend*, our heroine Gladys first notices her hero when his car pulls out of a garage:

> A black Mercedes S class model with tinted windows purred out. She stared at it in appreciation as it turned towards her, distracted from her quandary for the moment. She took a step back, startled, when the car stopped beside her. Her mouth almost gaped when the back passenger window rolled down. She couldn't help but note the looks of the man who stared back at her. He was very attractive, probably in his mid-thirties, with a certain stamp of authority all over him. His hair was bushier than she usually saw on men his age, but it was full and well-combed. The line between his deep set golden-brown eyes looked ingrained over the strong nose and pink lips accentuated by his dark skin. (Whitman 6)

Both car and man are made objects of desire by Gladys's appreciative gaze, a gaze reciprocated by him staring back at her. Although they exchange looks, though, it's not an even trade since he then scrutinizes (Whitman 7) her before offering her a lift, retaining his patriarchal powers of selection and control. She can find him attractive, but it's his scrutiny of her that decides their fate. Later, when he drives by and she wonders if he's seen her, we're told, "Edward

---

[10] Ifedigbo calls Whitman's novels homegrown, although the Nigerian-born Whitman now lives in diaspora in Seattle.

had seen Gladys alright. How couldn't he have done so when he'd deliberately altered his route to pass through this street again?" (Whitman 30).

When she gets into his car, having necessarily demurred first, she subordinates herself by making herself look away from him: "Gladys forced her gaze away and glanced around. The large sports car was spacious and lavishly appointed. She admired the shiny fixtures and automatic controls between the front bucket seats. The whole interior was fitted with smooth tan leather. She could imagine her car-loving brother's envy when she told him about it later" (Whitman 7). Turning her attention to the car's luxurious interior, her description lingers over the vehicle in the same way it lingered over the male body. As a prosthetic extension of the self, the car signals class, status, and success, all necessary matchmaking qualifiers.

It's noteworthy how often cars crop up when Jan Cohn lists the property inventories of romance fiction's desirable bachelors in her study *Romance and the Erotics of Property* (19n7, 46, 154). When Alhaji Abdu gets his comeuppance and loses everything in *Sin Is a Puppy*, foremost on his list of losses is his automobile: "He had no car; no stall; no merchandise; no money" (Yakubu 104). And, in keeping with *Americanah*'s deconstructionist approach to romance, the elusive love interest in that novel confesses early on that his acquisitions have begun to make him "feel bloated . . . the family, houses, the cars, the bank accounts" (Adichie 26). We see both Obinze in *Americanah* and Edward in *A Heart to Mend* being driven by hired drivers, and for Edward, his ability to afford this is part of what makes him desirable. A hero straight out of the Victorian mold, his name says it all—Edward Bestman is the best man for Gladys. He shares Edward Rochester's first name and Heathcliff's history of an orphaned child who makes good in the world. He holds out the promise of elevating Gladys's position through marriage in that same tradition— something the novel winks at in a moment of metanarrative: *"Dreamer*, her inner voice mocked, *you've read too many romance novels. This guy drives the latest car models and lives in the posh part of town. He's not going to notice you"* (Whitman 29).

The automobile accelerates the erotics of the text, including a latent homoerotics that the mainstream romance can't otherwise come to terms with, through the triangulation of desire as Gladys imagines her "car-loving brother's envy." In imagining this, she calibrates her own lust for us, as we learn that "her brother had gone crazy when she described that car. But she didn't tell him about the handsome owner who was more remarkable than his car" (Whitman 29). When the hero recalls their encounter, part of her appeal is that she admired his car; it's completely wrapped up with how she looks (both in the sense of her physical appearance and in the sense of her gaze): "The way her nose turned up at the tip; her delicate eyelids over large dark eyes; the wonder in her gaze as she had admired the car; the way her skirt had ridden up just so to reveal smooth, round knees, and give a hint of other shapely curves too. His heart rate hiked up in a pulsing beat of urgency—a marker of his physical attraction" (Whitman 18). Lending new meaning to the term autoerotic, the book stages love scenes in the car:

> She met him halfway as he bent his head to kiss her again. Her lips parted in welcome, even as he moved his hand around her ears to her collarbone and then lower. He caressed the top of her heaving breasts for a while and then slid his fingers beneath the bodice of the top to touch her breast. Her chest rose and fell as her breath came faster and when she gripped his biceps; his lower body came fully to life. He just had to feel those breasts against him and as he pressed her back into the bucket seat, she moaned against his lips. (Whitman 72)

The car's intimate space provides a literal vehicle in which the clichés of the genre can unfold.

Love

A world away from Lagos but still in Nigeria, the northern city of Kano is epicenter to the phenomenon of *littattafan soyayya* or love literature (also translated as books of love). Although these novels function inside the constraints of a strict Islamic culture, and therefore diverge in all sorts of ways from Mills & Boon–style romance novels, they nevertheless still share a conventional preoccupation with love and marriage. In her extremely useful discussion of Islamic feminism in northern Nigerian fiction, *Privately Empowered*, Shirin Edwin takes Novian Whitsitt's point that the novels mark a confrontation with dramatic social change while emphasizing that "the personal and private empoweringly serve as the motor for an entire socio-literary apparatus" (Edwin 28), rather than the public and the political.

However incidental or coincidental their word choice may be, Whitsitt calling the books "vehicles" for the writers' social concerns and Edwin talking about the "motor" for them being "quotidian private preoccupations" (Edwin 165) reminds us that these books also occupy a petrolandscape in which cars and the men who drive them feature prominently. This is amply illustrated in Glenna Gordon's photo-essay about the Kano authors, *Diagram of the Heart*, by an image of the cover of the love book *Mardiyya*: It has the author's name twice:

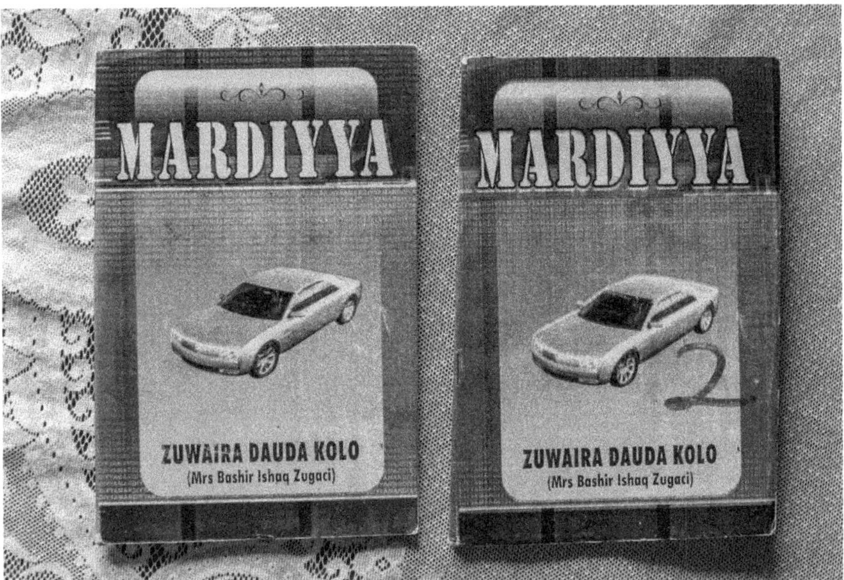

Figure 1. Side-by-side copies of the "love book," *Mardiyya*, on a lace tablecloth. From Glenna Gordon, *Diagram of the Heart*. Red Hook Editions, 2016.

"Zuwaira Dauda Kolo (Mrs Bashir Ishaq Zugaci)," with her married name included presumably for decorum, above which is a picture of a silver sedan, alone against a green backdrop (128–29). Gordon's photograph simply shows two side-by-side copies of the book on a lace tablecloth. Although the doubling of the image seems to double the import of the featured cars, the lace background immediately domesticates anything about the book that might otherwise have been associated with the outside world of automobility. Littattafan soyayya, a subgenre of Kano market literature (named for the marketplace where it is sold), has garnered some critical attention, but not much, if anything, has been said about the role oil plays in it. This is almost certainly because Nigeria's oil industry has, to date, operated solely in the South, although for years there has been speculation about whether northern wells show prospects of crude.[11] The

---

[11] Initial reports in 2016 were optimistic ("Good News: Oil Wells Discovered in Northern States of Nigeria," *Savid News*, Oct 15, 2016), followed by doubt ("$3b Down the Drain, Oil in North Remains Elusive," *The Guardian* [Nigeria], Nov 6, 2017), followed by renewed (if qualified) hope ("Excitement in Northern State as 7 Wells 'Confirm' Existence of Crude Oil," *Nigerian Bulletin*, May 25, 2018). By 2022, a state-owned Nigerian company had started drilling in the North (Onuah).

case for petrofeminism in the oil-rich (or oil-dependent, depending on your perspective) South of the country is more self-evident than applying the same term to the oil-poor and impoverished North, but the literature makes the case for itself—it is preoccupied with the automobile *despite* not being set in a place saturated by oil rather than because of it.

Sin Is a Puppy That Follows You Home by Balaraba Ramat Yakubu, originally published in 1990, was the first Hausa novel translated into English, in 2012. Concerned primarily with the female characters' matrimonial statuses, it gives us a number of related storylines, including that of a young woman, Saudatu, who aspires to be married before she finishes school. "'I don't trust men who like to give girls lifts in their cars,'" she says, and is told, "'You're right—you should be wary of the ones that want to give you a ride—they're all bastards. They just enjoy corrupting girls. Take a taxi, or walk, but never get into someone's car alone"(Yakubu, 29). Soon thereafter, in a scene reminiscent of *A Heart to Mend*, "an Alhaji drove by in a flashy gold-coloured car. They locked eyes for a moment, but she looked away, he passed her without slowing down. She saw him looking back at her, though, as he drove away" (Yakubu 30). Their exchange of glances preserves her modesty, which is reinforced when they meet again and he adjusts his rear-view mirror to steal looks at her without her knowledge or reciprocation, but suggests the start of sexual tension between them. In keeping with a context in which everything is commodified, including the women in the story who are constantly called "worthless," her interest in him is based entirely on his car, which "looked expensive and well-maintained" (Yakubu 53). This exaggerates Cohn's idea that the "heroine must not only be aware of the hero's display of consumer goods, she must see the hero in part *through* that display" (46). Luckily for her, he is both rich and kind (unlike some of the other men in the book, who really are bastards), and she gets to marry up out of poverty as his second wife.

In Edwin's discussion of *The Virtuous Woman* by Zaynab Alkali,[12] she describes a similar caution given to Nana, the protagonist of that book, who is told, "'Remember all the things I have been telling you about long journeys. . . . What did I say about accepting favours from strangers? . . . That includes free car rides, monetary gifts and clothes'" (Alkali qtd Edwin, 122). Edwin reads the book, a deliberately moralistic novel intended for adolescents (Alkali qtd Edwin, 116), for its African-Islamic feminism, which comes to the fore especially when Nana declines a lift from a stranger. Edwin enumerates the qualities that "comprise strong feminism," including Nana's courage, strength of

---

[12] Alkali, Zaynab. *The Virtuous Woman* (Longman, 1987).

character, decisiveness, and dignity, all of which happen to be made visible by "her steadfast refusal of a free ride" (123). She makes this choice on day one of a journey that lasts the whole book that others have read as allegorical[13] but that might equally be taken at face-value for its concern with transit and everything that goes along with it (from a government-appointed travel escort to the bus stopping for prayers to conversations with fellow train passengers). It is specifically an empowering journey, toward school and a scholarship—not just feminist but petrofeminist in its enabling mobility.

Time magazine profiling one of Glenna Gordon's authors as "Kano's Jane Austen" might initially sound like a sweeping temporal and historical conflation, but when a different author featured in Gordon's book says she "loves Jane Austen novels" (20) we need to understand that a certain cross-cultural identification is in progress having to do with the circumstances of their literary production. Austen's world made no room for women writers, and the women writers of Kano, according to Gordon's introduction to *Diagram of the Heart*, are not supposed to leave their houses (15). Spatially constrained, they nevertheless let their imaginations loose. They also share modes of production across time and space—handwriting (in small composition books, in the case of the Nigerian writers) and self-publishing. Despite laments that the book in Nigeria is an endangered artifact (recall that "Nigeria is a bookless country"), the technology of the book holds its own against the digital technology of the mobile phone, which we can see in the final image in Gordon's book, a Hausa woman reading a romance novel on the train from Lagos to Kano. In transit, she reads a hard copy of the book by the light of her mobile phone, harnessing the newer technology in service to the old (130–31).

In her essay "Reading Romance Novels in Postcolonial India," Jyoti Puri affords us another cross-cultural comparison[14] when she describes young Indian women's negotiated reading habits, including ignoring remonstrations from parents, delaying reading the material, and concealing their consumption. One such young woman, "Reshma, who loves 'intense, meaningful, hot, sexy stories,' says of her parents, 'They don't know.' She reads the novels in the

---

[13] "*The Virtuous Woman* is a political allegory as well as a parable about gender. The three lorries are Nigeria's three regions, rushing headlong toward disaster in the early 1960s, while the women-and-men plot suggests a womanist healing of Nigerian divisions" (Griswold, 184).

[14] The degrees of cultural separation here may be even fewer than one would imagine: Shirin Edwin pointed out to me in conversation that it's no coincidence that *Sin Is a Puppy* was published by an Indian house—it's the kind of conventional narrative that appeals to a mainstream Indian audience, she says. Moreover, as she explains in *Privately Empowered*, the books themselves are heavily influenced by Bollywood films (26).

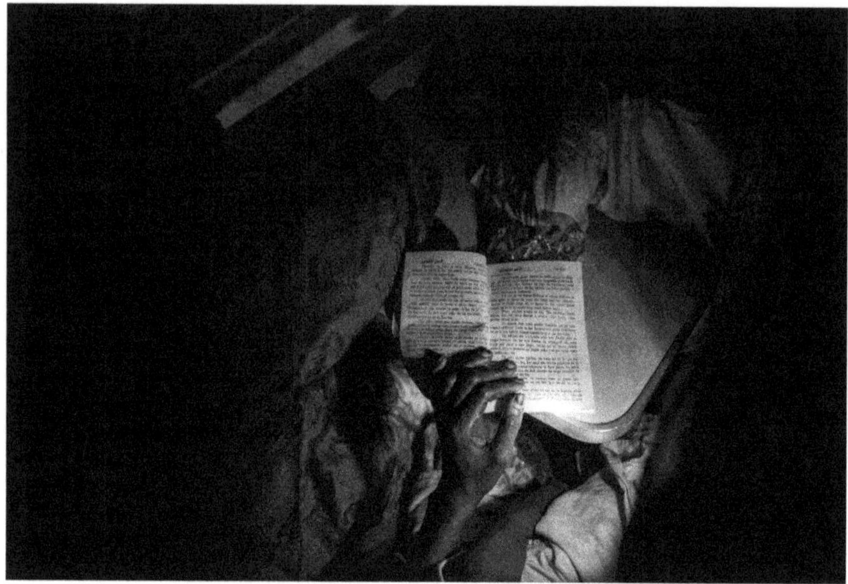

Figure 2. Woman reading on train by the light of her mobile phone. From Glenna Gordon, *Diagram of the Heart*. Red Hook Editions, 2016.

bathroom, in the train, and in college" (440). Puri contemplates whether the act of reading romance novels can be considered cultural resistance, a challenge to the hegemonic order, but concludes that "the act of reading is limited as a political strategy" since it "remains an isolated, individualized activity" and "may contain and neutralize women's discomfort with their realities" (441–42). However, to return to Radway's phrase, the event of reading matters here. To read in liminal spaces, like on the train, is to practice and perhaps even to model behaviors not sanctioned by mainstream authorities, who, in the case of the northern Nigerian authors, include the morality police and the Ministry of Education as well as parents and spouses. In addition, the Kano women's writers' cooperative confounds the assumption that reading and writing are isolated activities.

Working It

In Ann Rosalind Jones's essay on Mills & Boon, she argues that although the demands and debates associated with feminism produce striking ambivalence in these novels, she does see real innovation, over time, even if, ultimately, those changes contain feminism's radical potential and don't negate the basic premise of the genre—that is, that the greatest goal and pleasure in a woman's life is the love of a good man (211). One of these changes is a heroine's commitment to her work, and interestingly, Jones gives the example of a Harlequin romance called *Maelstrom*, in which both hero and heroine are petroleum engineers who risk life and limb to cap an exploding well. Novian Whitsitt's study of love literature lets us trace in it a parallel trend "within a working paradigm of African feminism" (142), specifically an Islamic-Hausa breed of feminism, of promoting the working woman who can "fall in love and lead a blessed life of motherhood, career, and opulence beyond her expectations" (147).

Dirt, like work, is traditionally gendered, so that if women's dirt is household dust, the dirt of oil is usually associated with masculinity. Thinking about petroleum through the lens of feminism, however, lets us see the dirty work of oil as being as much women's work as men's. This is true not only at the level of impoverished Nigerian women siphoning oil at the pipeline to sell by the side of the road, but also in the highest ranks of the industry. Although, in the words of the BBC, the Nigerian oil industry is still a "boys' club" (Orin Gordon), there is a growing number of female oil executives in Nigeria, part of an indigenization plan for the industry. Figures like Folorunsho Alakija tend to make the news. According to *Forbes*,

> Alakaija is worth a staggering $1.73 billion . . . making her the fourth richest person in Nigeria and second richest woman in Africa. She is the Vice Chair of Nigerian oil exploration company, Famfa Oil, which shares a joint partnership agreement with international giants Chevron and Petrobras. With a 60 per cent stake of block OML 127 of the Agbami field, one of Nigeria's largest deepwater discoveries, Famfa Oil produces approximately 250,000 barrels of crude per day. ("A Head for Business")

It's no coincidence that in *A Heart to Mend*, Gladys works hard to qualify for a position with Zenon Oil, a choice reflecting the country's total immersion in the oil industry, a bigger profile for women in it, and the romance novel's increasing interest in women's work.

Corruption and scandal run rampant in Nigerian society, with some of the newly high-profile women in oil ensnared in this other sort of dirty work, such as former petroleum minister Diezani Alison-Madueke who was charged with money laundering in a case involving the misappropriation of funds from the Nigerian national oil company NNPC in a bribery scandal to keep then-president Goodluck Jonathan in office. This corruption amply illustrates "the ongoing romance between unanswerable corporations and unspeakable regimes" (105), to recall Rob Nixon's line. Nixon's choice of the word *romance* implies an unsavory relationship but also speaks to a secondary sense of the word, that is, "an extravagant fabrication; a wild falsehood, a fantasy" ("romance"). Any work of romance fiction is a fantasy, "remote from everyday life" ("romance"), but this meaning rings especially true when the promise of sentimental love is premised on a romantic scam.

Slickness

The subject of two *Forbes Africa* profiles, Alakaija is described as a "Slick oil baroness" on one of the magazine's covers, with the punny modifier "slick" introducing a query the accompanying article does not ask—to what extent is she slick because she's a smooth operator and to what extent is she slick because she's a swindler, having greased palms along the way and benefited from corruption at the highest levels? Regardless, slickness, with all its connotations, works. One iteration of Nigeria's infamous scamming economy is the romance scam, the new 419 (so called after Article 419 of the country's criminal code, having to do with fraud) (Koerner), in which scammers frequently pose as oil rig workers since it's a profession that covers a multitude of sins, including absenteeism, lack of communication, and the need for funds. *Wired* magazine calls it a "macho cover story that allows them to fade in and out of victims' lives at will" (Koerner). A microcosm of the global scam that is Big Oil, these smaller scams represent the latest installment in the story of leveraging oil for profit at the expense of the vulnerable—in this case, those most susceptible to romance. Cyberpsychologists say that women who fall for online romance scams "tend to score highly on tests that measure how much they idealize romantic love" (Koerner). A set of guidelines available online about how to avoid being scammed flags "warning signs on the road to romance," which might be ""hazardous" or "slippery" (Business Wire). But the scammer's slipperiness or slickness is also his appeal.

Accounts of romance scams read like mass market paperbacks. Describing how one woman was taken for a ride, the *Wired* article gushes in pink prose, "Mike Benson" was "a dashing oil worker to whom she'd sent around $14,000 over the preceding months" (Koerner). Another scammer, posing as Duke McGregor, claimed to be a mechanical engineer with Transocean offshore drilling contractors: "When he wasn't working on North Sea oil rigs, he enjoyed reading classic novels, playing with his tiger-striped tabby cat, and strumming a heart-shaped guitar." Duke's handsome profile photo allegedly showed a middle-aged man with a ruddy face, strong black eyebrows, and a welcoming gaze (Koerner). The victims of these scams, we are told, often share a particular psychological trait: an exceptional faith in the existence and importance of romantic destiny (Koerner), which is, of course, also the reason why women read romances.

Drilling Down

I began this chapter by referring to Hitchcock's law that oil is "everywhere and obvious, it must be opaque or otherwise fantastic." Both of these things seem to be simultaneously true for romance novels, in that the fictions are first and foremost predicated on an enabling, mobilizing, largely invisible petroculture, and, that secondly, when oil is visible, it is fantastically so, as in the rush to cap the gusher in *Maelstrom*.

In "Petrofiction," his 1992 trailblazing essay, Amitav Ghosh observes that "the history of oil is a matter of embarrassment verging on the unspeakable, the pornographic" (29). The pornographic qualities of oil extend to an interest in the money shot. Textual representations of oil gushers, such as scenes in Upton Sinclair's 1927 novel *Oil!* or its 2007 cinematic remake *There Will Be Blood*, have been read as adolescent ejaculation fantasies (LeMenager, "Aesthetics" 74), where the spectacle of the blowout occludes the slow violence (76–77) of the oil ordinary. The same would appear to be true of a similar moment in the romance canon; according to Ann Rosalind Jones, *Maelstrom*'s "crisis is genuinely exciting: the couple risks life and limb to cap an exploding well, a spectacular accomplishment" (211). The relationship, however, between romance and porn (and therefore, by extension, between romance and oil), is highly contested. In 1979, Ann Barr Snitow famously equated romance novels with porn for women, arguing that because they are formulaic, "the novels have no plot in the usual sense" (309). A study in the *Review of General Psychology* makes virtually the opposite argument:

> Unlike romance novels, pornography does not really have a "plot." Instead, they typically contain a loosely connected series of sex scenes, each of which usually ends with a visible ejaculation, the "money shot." As a result, pornography has as many climaxes as it does scenes (at least). A romance novel has only one climax, the moment when the hero and the heroine declare their mutual love for one another. (Salmon 155)

Another study, entitled, "'She Exploded into a Million Pieces': A Qualitative and Quantitative Analysis of Orgasms in Contemporary Romance Novels," published in the journal *Sexuality and Culture*, has it that "female characters were significantly more likely to be depicted experiencing an orgasm during a sexual encounter than male characters" (Cabrera and Ménard 199). Perhaps the difference lies in degrees of visibility—the money shot and the geyser show spectacularly while the female orgasm and the everyday backdrop of a petro-culture usually do not. Maybe the difference is one of degrees of economic conventionality—the romance novel is, fundamentally, a fetishistic capitalist fantasy that must play out within the monetary mainstream to be fulfilling. Anne Cranny-Francis puts it this way: "Romantic fiction is the most difficult genre to subvert because it encodes the most coherent inflection of the discourses of gender, class and race constitutive of the contemporary social order; it encodes the bourgeois fairy-tale" (192).

That bourgeois fairy-tale pertains regardless of setting; every study of romance fiction makes allusion at some point to its formulaic quality (e.g., "The Eight Essential Elements of the Romance Novel"[15]), which is why the romance novel as genre will never entirely spill over from readerly pleasure to writerly *jouissance*. The single constant is not the attraction, the barrier, or the declaration but the highly conventional pursuit of financial bliss. Everything else—the launching of a career, the happy ending, the heterosexual marriage, the promise of offspring—sublimates itself when an economic reading is foregrounded. Thus, Hallemeier reads *Americanah* as about the "pursuit of capital and love alike" (242): "*Americanah* presents an alternative, utopic vision of global power in which the United States stands as a foil to the promising future of late Nigerian capitalism" (232). Presumably in this dream of Nigerian capitalism (237), oil would be part of the success story, but Hallemeier gets through her entire essay about capital in the novel without mentioning oil or petroleum. Oil as generative backdrop is so assumed that neither critics nor characters see it. It does not perform itself; it is not pornographically visible. It is sublimated in the same way that women's reproductive labor is sublimated.

There is a moment late in *Americanah* when Obinze warns that oil companies aren't doing Nigerians any favors (Adichie 580), but his success in real estate is predicated on renting to oil men. Everyone in the room at this particular party has made a fortune from some aspect of the petrostate: Eze, "the wealthiest man in the room," is "an owner of oil wells" (577) and "Ahmed had leased strategic rooftops in Lagos just as the mobile phone companies were coming in, and now he sublet the rooftops for their base stations and made what he wryly referred to as the only clean easy money in the country" (576). Obinze's fictional warning about "big oils" planning to move offshore, leaving

---

[15] Pamela Regis, *A Natural History of the Romance Novel* (U of Pennsylvania P, 2013).

onshore operations to the Chinese, is framed by the financial press in real life as an indigenous opportunity.[16] *Homegrown*, a word I used at the top of this chapter, may seem an inapt descriptor for oil (better suited to, say, agriculture), but it is invoked with some frequency to describe a new phase in Nigerian oil production, one in which local entrepreneurs take over operations. In the dream economy Adichie's characters aspire to live in, domestic oil control is the material correlate of the textual development that is the homegrown Nigerian romance novel.

Love in *Americanah* does cross boundaries, breach borders, and challenge norms (Leetsch), but it does so within a conventional economic system and is fundamentally conservative in its interest in "manifesting capitalism" (Hallemeier 235). This is even more true of the other romance novels read here, although they, while more traditional, may actually have more radical potential. *Sin Is a Puppy* and other love books, for instance, promote female education and have been credited with a rise in literacy among Hausa women (Khanna vi).[17] They are also much more likely to meet with censorship and public criticism, according to Whitsitt. Romance as a genre tends to play it safe, and these feminist fantasies are predicated on oil futures that create a safety net for women's welfare, well-being, and ultimately on being well-off. Petrofeminism operates within, not outside of, an oil economy, making space inside it for women to claim what is rightfully theirs without having to siphon it off riskily, a space where the romance of women's work is a job at the oil company, and where a romantic relationship is not possible until the love interest returns to Nigeria to make his fortune renting properties to Shell. Green-Simms remarks on similar constraints to a feminist automobility: "In a society where women's autonomy is constrained and limited by patriarchal regimes, the car is a machine that can, in certain cases, instantiate new sources of power and pleasure" (*Postcolonial Automobility* 191). Read alongside each other, the romance stories of northern and southern Nigeria offer glimpses into competing modernities, quite different in some essential ways, and yet both are products of a petroculture shaped and scribed by women. We can read for romance elsewhere too: in short stories by writers like Sefi Atta and Nnedi Okorafor, which demonstrate love and care for nation, community, and environment. As discussed in Chapter 1, those short stories explicitly

---

[16] See, for example, "Meet the New Face of Nigeria's Oil Industry" (Jeff Grey, *The Globe and Mail*, 10 Aug. 2014).

[17] See Edwin on this, too.

sabotage the status quo, and while the romance novels do not, playing instead within received forms of genre and capital, the Nigerian love story does do "other things," per the sentiment of this chapter's epigraph. However we understand the petro-imaginary, it must ultimately make all of these iterations visible, and petrofeminism lets us see the inseparable nature of women's politics from oil.

CHAPTER 3

PETROART

A Thing That Seeps

The product we know today as *petroleum* gets named in the Middle Ages. Apocryphally, sources tell us again and again that the word, Latin for "rock oil," was popularized by Georgius Agricola in his 1546 book *De Natura Fossilium*,[1] but in an intense dive into its actual origins, Grantley McDonald painstakingly clears away the etymological cobwebs to show us that, while "Agricola drew on a rich tradition of employing petroleum in medicine," he did not coin the word. Instead, McDonald says, the honor should go to Constantinus Africanus, the Saracen physician of the eleventh century (363). In other words, it is Africa that names the thing we call petroleum.[2]

The contemporary sculptor Sokari Douglas Camp, born in 1958 in Nigeria and based in London, creates what we might call *petroart*[3] with that complex etymology in mind. Like the word *petroleum* itself, her sculptures reflect a tangled and contested history of laying claim to oil. While her works often recuperate and reappropriate Western art for African purposes, they do much more than that: They reference and then redirect preexisting connections to oil histories, opening up a pipeline between the West and what the West has traditionally thought of as the rest.

---

[1] See the "Petroleum" entry in Wikipedia, among other places: https://en.wikipedia.org/wiki/Petroleum.

[2] Of course, human/oil relations date to far earlier than the modern "discovery" of oil, as an energy source, among other uses. See, for instance, "History of Oil" EKT Interactive, 2004. Web. 5 April 2024. (https://ektinteractive.com/history-of-oil/#:~:text=The%20first%20oil%20had%20actually,for%20the%20new%20oil%20economy).

[3] Not to be confused with PetroART (Petroleum Advanced Research and Technologies), Saudi Arabia's center for petroleum research and development.

Douglas Camp does this squarely within the tradition that undoes tradition—postcolonial practice. As Bill Ashcroft, Gareth Griffiths, and Helen Tiffin remind us about how colonial power worked in imperial parts,

> it is impossible to discuss the culture of such societies without recognizing the power of colonial institutions, ideologies and patronage systems in validating some forms of culture and denying the validity of others. The privileging of writing and other inscriptive arts over the oral and the performative arts, as well as over other kinds of signifying practices such as sculpture, painting, carving, weaving, ceramics—the whole body of material inscription beyond the written—offers a classic example of such privileging. (52–53)

By picking sculpture as her medium, Douglas Camp challenges that index of privilege—in her art, the empire does not write back, it sculpts back (Schlote 256).

"An oil barrel starts this conversation," writes Douglas Camp (Stoa 169). When asked, she says it is a conversation she is holding with herself (personal interview), but it is also clearly a broader one about our place in the material world. She describes her process as "welding, cutting, and bending sheet steel & recycled oil barrels into shape," which suggests they are not in the right shape to begin with. Reshaping this emblem of the modern age of oil extraction, containment, and transport allows for reframing the conversation about those practices. When Douglas Camp employs oil barrels in pieces that echo "classical" works of art, like Sandro Botticelli's[4] *Primavera* (c. 1482), it serves as a way to visually tell other stories of oil's forelives, rerouting the narrative through canon, and, in so doing, disrupting both the usual story of oil and the canon itself.

Blind Love and Grace (2016) incorporates steel, oil barrels, aluminum, gold and copper leaf, and acrylic paint to reinterpret the central figures in Botticelli's *Primavera*: Venus beneath a blindfolded Cupid. In her new context, Venus, cupped by steel banana leaves, summons growth from the oil barrel. The piece challenges the viewer's assumptions across categories: It is anticanonical and yet it relies on the canon for meaning. It translates between forms, from painting to three-dimensional object, so that the figures unfold from the wooden panel to take up space assertively. It elides the industrial and the natural, remaking petro-objects to achieve that elision.

The Shell Transport and Trading Company, a precursor to the company we know today as Shell, was named in honor of its founder's father who made

---

[4] Delightfully (but coincidentally) Botticelli's last name was first a nickname meaning "little barrel."

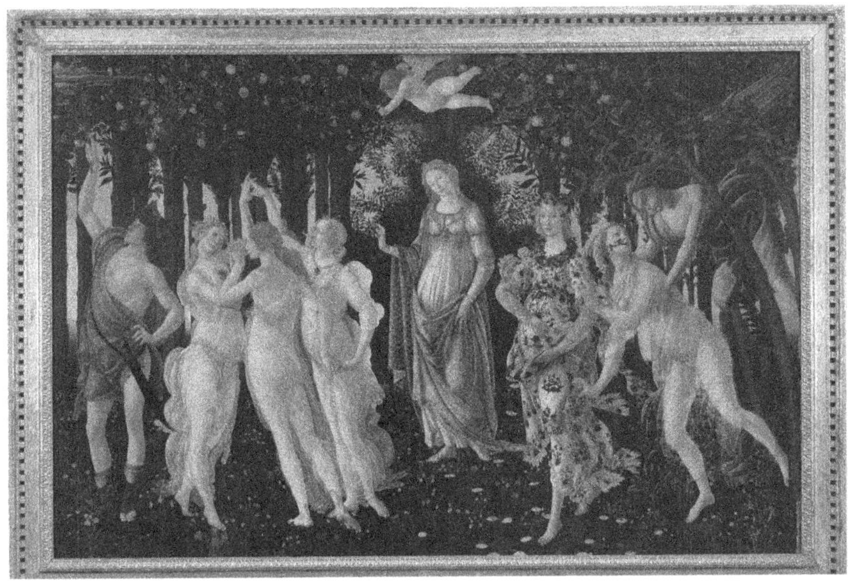

Figure 3. Sandro Botticelli, *Primavera*. c. 1480. Tempera grassa on wood. The Uffizi.

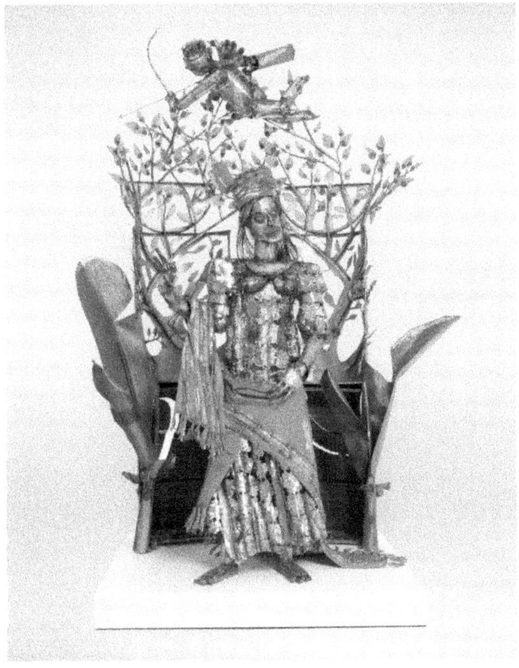

Figure 4. "Grace" stands between oil barrels cut in half in Sokari Douglas Camp's *Blind Love and Grace* (2016). © 2024 Artists Rights Society (ARS), New York / DACS, London.

his fortune as a "shell dealer and importer" (Barnett 247) on the coattails of the Victorian era's "shell fever" (245). Although its original logo was "a bland mussel, "in 1904, Shell introduced its iconic ribbed scallop, so recognizable that it no longer requires the company name" (261). In what at first might strike us as a wonderfully anachronistic analysis of *Primavera*, Douglas Camp reads contemporary corporate semiotics into the fifteenth century painting, connecting Botticelli's design to the Shell logo:

> The extraordinary thing about the "Primavera" work was to discover that there was a little element of Shell's logo in that work. Right in the middle of the painting, the forest starts to break up and it's a sort of halo around one of the central figures . . . the forest is there, opening up a bit like a shell. And Botticelli . . . did Venus coming out of a shell. The shell, and the landscape breaking up because something is so beautiful, a shell carrying a beauty—all these logos were used by Botticelli and then the designers of Shell, the petroleum company, used Venus's shell as their logo. (Douglas Camp)

As it turns out, however, conceiving of a Renaissance logo is entirely in keeping with those times. *Primavera* is replete with symbolism, including five hundred identified plants (Capretti 49), creating a complex meaning-making system some have called a "coded language" ("Characters"), in the same way that modern logos encode brand connotations. Also, as a logo or slogan would carry over from one advertisement to the next, so too Venus and her shell repeat. Although *Primavera* was painted before *The Birth of Venus*, Venus's story happens the other way around: She comes ashore, soon to be modestly covered in myrtle, the plant with which she becomes associated in her new role as goddess of marriage in *Primavera*. In *Blind Love and Grace*, Douglas Camp's sculpture, the oil barrels cut in half that open up to reveal Grace clearly evoke Venus's shell at the same time as nodding directly to *Primavera*. In the painting, silhouetted tree branches form the "halo" of negative space Douglas Camp describes. In her sculpture, these turn into structural elements, holding Cupid aloft. Thus Douglas Camp conflates the two Botticelli paintings, triangulating them through her own work via the Shell/shell iconography. The violence that comes with Shell oil then enters into the paintings, breaking up the landscape under the sign of beauty, just as it does in the Niger Delta.

In his essay for the exhibit catalogue from London's October Gallery, Gerard Houghton speaks generically about how Douglas Camp "Africanises" the works she "reprises," but the sartorial choices she makes for her figures are self-evidently Nigerian ones ("C.B.E." 6–7). In *Blind Love and Grace*, Venus's

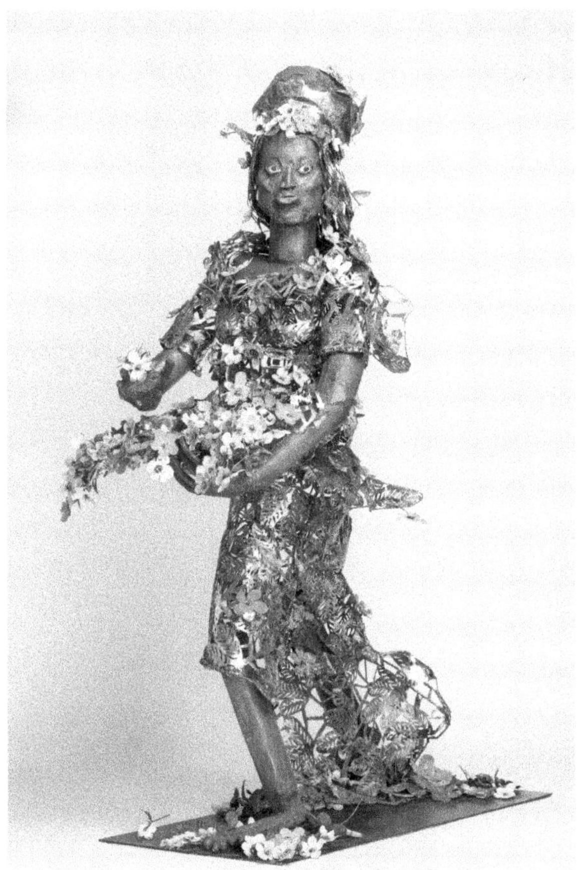

Figure 5. Note the toy cars scattered across Flora in Sokari Douglas Camp's *Primavera* (2015). © 2024 Artists Rights Society (ARS), New York / DACS, London.

red shawl turns into a distinctive Kalabari hat and wrap, worn by that eastern Niger Delta group. In *Primavera* (2015), she borrows Botticelli's title and another of his subjects from that painting, this time reimagining Flora in a *gele* headdress, decorated from head to toe with steel flowers and toy cars. As Houghton notes,

> The importance of cloth fabrics in denoting status and wealth is common throughout West Africa, and Kalabari culture puts particular emphasis on its value as the significant vehicle of cultural transmission. Sokari's figures dressed in identifying clothing are less vulnerable to misinterpretation than their predecessors, and in her modernised version . . . subtle shifts

have been introduced to imply that the imbalance and exploitation of former times is at last, finally beginning to be redressed. ("C.B.E." 6)

But Houghton misses the double entendre in his own wording: Douglas Camp's work redresses past wrongs by *re-dressing* these received characters in particular cloth, clothing, and costumery and thereby "sculpting a history that is never talked about—a Black cultural history" ("The Fabric of the World" ["FW"] 13). Dressing, with its scope for cross-dressing across cultures, genders, and temporalities, enables a combinatory conceptual process: "Because of my own interest in dressing and performance, specific costume elements . . . inspired me. As a Kalabari woman I enjoy the experience of dressing; there's so much information there to play with. So, I feel my way forwards instinctively, dressing each layer upon another, to achieve some new synthesis" (Houghton, "FW" 7). This experiment in hybridity is not what Houghton warns of when he says, of her cars and flowers combined, "If this were a Greek myth, such hybridity, indiscriminately scattered about, would foretell the advent of peculiar fruit and troubling consequences" ("C.B.E." 7). For Douglas Camp, who thinks of herself as "stamped by the triangle" (Houghton, "FW" 13) of Nigeria, Britain, and the Caribbean, hybridity arises from diasporic histories and inheritances.

Fabric recurs thematically throughout Douglas Camp's work, both as tangible textile and as abstraction about hybridity and the ways in which we are woven together (Houghton, "FW" 14). A fabric, as opposed to a cloth, is a manufactured material, and its derivation from *faber*, a worker in metal, stone, or wood ("fabric"), sums up the effort of fabrication that goes into creating the illusion of seeing one material as another. Take, for instance, *Primavera*, with its "heavily embroidered African 'lace' robe—itself an exercise in skilful working with a plasma cutter" (Houghton, "C.B.E." 7). It is especially in regard to the figuring of fabric in steel that critics remark on the suspension of time in these sculptures, noting that "Douglas Camp has managed to imbue the industrial material with a sense of weightless motion" (A. Young, "Sokari Douglas Camp") and commenting on her "ability to turn hard steel into flowing fabrics, feathers or even flesh and bone" (Houghton, "FW" 14).

Both Botticelli and Douglas Camp freeze the action in their *Primaveras*. The original painting captures a moment of transition: the nymph Chloris transforming into Flora, goddess of spring, after having been kidnapped by and married to Zephyrus. Botticelli achieves this by overlapping the two women on the right side of the composition, showing transfiguration in process in a still image. Likened to an ancient frieze, the painting holds its subjects in a moment of potential energy. Houghton emphasizes the movement stored up in it:

> The scene appears paused in a timeless moment of stillness before the
> arrow flies, and yet, at the same time, it swirls with an inner dynamic that
> flows from right to left. Movement begins from the irruption into the scene
> of the flying Zephyros—the cold West wind—who leans down to capture
> the attention of the nymph, Chloris, Goddess of Flowers, who is destined
> to become his wife. (Houghton, "C.B.E." 3)

Douglas Camp's Flora steps forward in the same posture as Botticelli's, but,
permanently adhered to the steel plate that anchors the sculpture, she is forever caught in the motion of "collecting or giving away" (*Primavera*) her flowers
and cars. Whether she gives or takes, she has the agency in this arrangement.
In the hands of nature's emissary, car culture—the ultimate realization of our
petromodernity—is miniaturized, calcified, and demoted to a plaything.

Douglas Camp picks up on the "fine clothes wealth and fantasy" (in her own
words) of the painting, reifying that into a "materialism" in which the abundance of spring springs forth in a mélange of nature and culture (*Primavera*).
However, there is no mistaking the artwork for the real of nature. By actively resisting collapsing nature into its representation, it avoids what Kevin
Hutchings warns about the practice of ecocriticism: that "by grounding criticism in an urgent sense of concern for the health of the natural world that
sustains us, such claims implicitly 'naturalize' ecocriticism, yoking its theoretical and practical imperatives to the material world itself and, in the process,
subtly downplaying its own status as a human discursive practice" (4). Flora's
fanciful gesture and her improbable bouquet de-yoke Douglas Camp's ecocriticism from practicalities. Houghton weighs possible interpretations: "Perhaps
new cars do represent the flowering tip of our modern technological civilisa-
tion. But there's a deeper note of ecological unease underlying Sokari's world-
view evident, here, in the erosion of distinction between objects of the natural
world and those of unnatural genesis" ("C.B.E" 7). Other critics also pick up on
that "note of ecological unease": "it is as if the wild, blooming landscape that
Flora represents has been invaded by a sprawling junkyard" (A. Young, "Sokari
Douglas Camp"). These readings presume an originary distinction between the
un/natural, however, whereas what Douglas Camp's steelwork instead regis-
ters is the always-already existing mutual constitution of nature and culture,[5]

---

[5] We understand this, and the central role of mediating discursive practice, from Raymond
Williams's *Keywords*, in which he demonstrates how the word *culture* has its roots in nature,
as in to cultivate or tend (87), and that nature is in turn associated with "what man has not
made" (223).

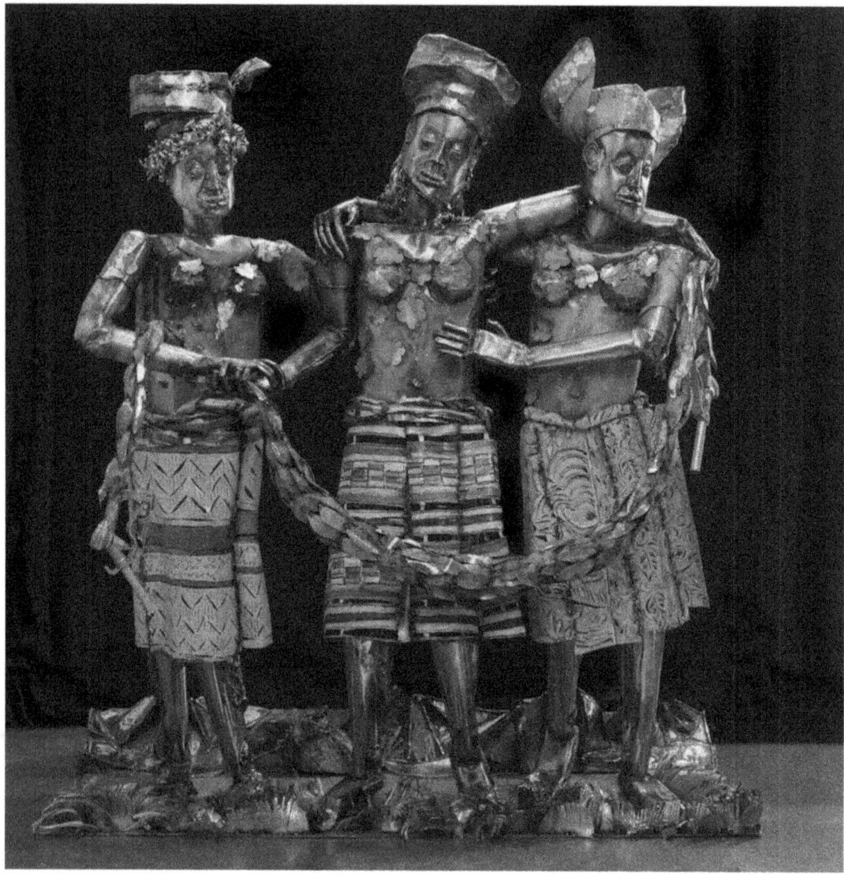

Figure 6. Three figures based on the William Blake work by the same name. Note the petrol nozzles at the ends of their garland. Sokari Douglas Camp, *Europe Supported by Africa and America* (2015). © 2024 Artists Rights Society (ARS), New York / DACS, London.

especially for the oil-bound Nigerian woman for whom these relations are binding and inextricable.[6]

In her version of *Europe Supported by Africa and America*, Douglas Camp takes another work from the heart of the Western canon, William Blake's 1796 drawing by the same title, and recenters it around that Nigerian woman and

---

[6] This is what Jennifer Price concludes about plastic pink flamingos: "Very literally, the plastic pink flamingo is wholly real, and certifiably natural. It's just nature that's been mined, harvested, sold, heated, boxed, resold, and reshipped" (87). In the same way that the lawn ornament (a petroleum product, of course) is natural, so too is the oil barrel, the model car, and the sheet of steel.

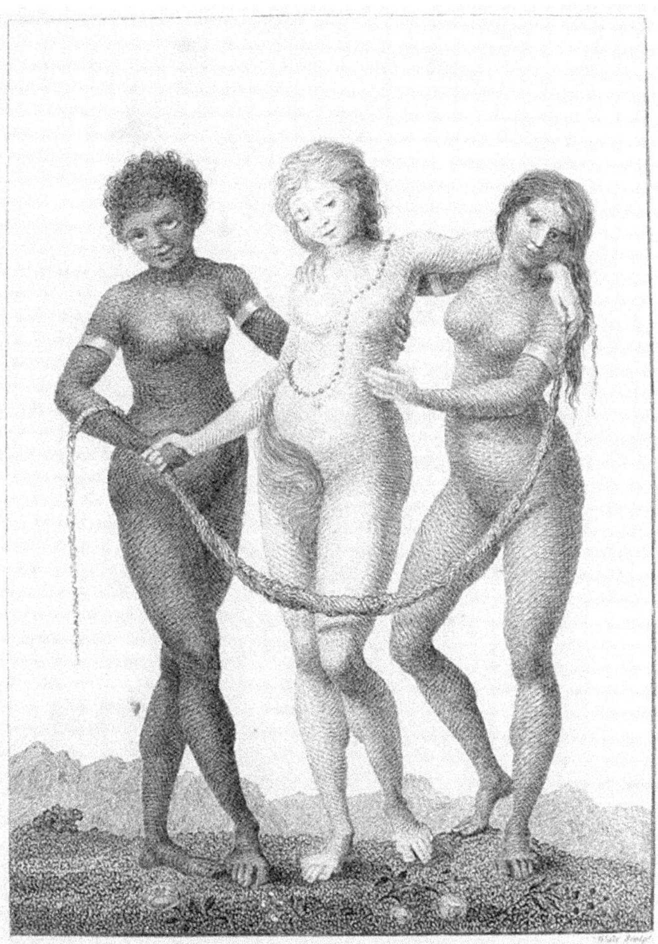

Figure 7. Female figures as continents—the inspiration for Sokari Douglas Camp's work by the same name. William Blake, *Europe Supported by Africa and America*. 1796. Print. © Victoria and Albert Museum, London.

her relationship to oil. Playing with artistic inheritance, and her place in it, Douglas Camp draws an aesthetic lineage from Botticelli to Blake: "His three figures were an echo from Botticelli's time" (personal interview). Blake's print, in the Victoria and Albert (V&A) collection in London, can be viewed online, where beneath the image appears the legend, "This object, or the text that describes it, is deemed offensive and discriminatory. We are committed to improving our records, and work is ongoing." This disclaimer mediates our encounter with the art and hints at something being not quite right, although we are not told what for sure. Possibly what's offensive is that their inequality is evident (signaled through the supporting women's armbands and their more exposed nudity). As the blurb explains, though, Blake's illustration appeared in an abolitionist text, John Gabriel Stedman's *Narrative of a Five Years Expedition Against the Revolted Negroes of Surinam*, and he intended its symbolism to suggest the ways in which the two Black women have been made to support the central white woman because of their enslavement. That was the point—to show a readership that needed to be convinced of the moral harms of slavery that people were being treated differently. Alternatively, it's the use of the word "negro" that offends, so the site finds it necessary to explain that the term "was used historically to describe people of black African heritage but, since the 1960s, has fallen from usage and, increasingly, is considered offensive. The term is repeated here in its original historical context." (Never mind that we encounter it through the intermediary online platform.)

Although scholars such as Anne Mellor read "racism and sexism" (357) into Blake's print, others, like Saree Makdisi, argue that, "On the contrary, the image could easily be read instead as a critique of a world system based on inequality and brutal exploitation" (256). This take aligns with what appealed to Douglas Camp about Blake in the first instance, realizing that "he was a campaigner in his own way, because he was involved in the abolition of slavery" (personal interview). In *Europe Supported by Africa and America*, Blake calculatedly subverts the master narrative. True, "Europe" is supported by the two other (also female, nubile) continents, but they are equally bound together, by the vine or garland they hold that emblematizes their naturalized relations but that also serves as a chain.[7] Moreover, "Africa" and "America" return our gaze while "Europe" demurely lowers her eyes. Although this could impute a brassy, brazen quality to them and an inherent virginal, virtuous quality to her, in the context of Stedman's reformist narrative, it instead suggests "Europe" averting her gaze from what the others confront directly.

Further down the V&A webpage is a link to Douglas Camp's version exhibited there in 2022–23, in which "Blake's graces are transfigured—she dresses them, and they stand united, equal in stature, adornments and attire" (Blake, *Europe Supported*). As in Blake's depiction, Douglas Camp's three allegorical figures also hold a verdant garland, only theirs morphs into fuel pump nozzles, like those you would use at a petrol station, replacing resistance and solidarity with commerce and oil as the ties that bind. "Business as usual?" Douglas Camp

---

[7] If the garland is made of tobacco leaves, as many have surmised, its association with plantation economies would increase its binding powers.

asks on her artist's page for the piece, suggesting that "the coercion in working with each other and not caring for the environment" (Douglas Camp, *Europe Supported*) dates back as least as far as Blake's print.

Douglas Camp sets out to decolonize the classical and to find a new, concrete form for marking where oil touches the human. To do this, her work embraces some core ambiguities, starting with her own equivocal relationship to the "original" pieces: She sets herself a fundamentally impossible task in her re-creations of these famous, touchstone works in that hers are perpetually referential. This has at least two implications: that her deconstructions can only be fully understood if her audience is familiar with the piece to which hers refer, and that the meaning she makes thus depends on a reappropriative relationship. Politically, then, "the focus on environmental conditions in Niger Delta" and her "colonialism criticism" (Douglas Camp, *Green Leaf*) must function within a set of Western referents. Working within that construct, though, also allows for its exploitation. The current asking price for Douglas Camp's "Europe" is €100,000. At the bottom of each work's page on her website we're prompted to consider a purchase: "Interested in acquiring this work?" Making a living making anticapitalist art looks a lot like reparations for "Africa" having supported "Europe" and "America."

Throughout, Douglas Camp retains an ironically playful stance about the place of petroleum in our world. The descriptive notes for *Green Leaf Barrel* (2014) and *All That Glitters* (2013) both emphasize their humor and optimism. The two pieces bear a family resemblance: They feature an oil barrel split in half with a woman, also made of metal, on top. In *Green Leaf Barrel*, she's kneeling to cultivate the Perspex plant growing out of the barrel. In *All That Glitters*, she's suspended above the sliced barrel "showing off being invincible" (Douglas Camp, *Green Leaf, All*). Together, they convey a conquering spirit, able to create nature out of the capitalist artifact and to ride astride it, though of course among the ironies here is the fact that not only is this nature artificial, but that Perspex, a branded form of plexiglass, is itself a petroleum product. Steel, too, is a problematic medium, a reality Douglas Camp acknowledges, calling it, in an interview with *Voice* magazine, "a very polluting material," and remarking that "the pieces that I've made about the situation in the Niger Delta have tried to highlight action and positivity, but are also made from steel and part of the destruction and deterioration of our environment" (James).

On her "about the artist" page, Douglas Camp shares the genesis of her use of oil barrels and her awareness of the utter ubiquity of oil:

Living in London and being gifted oil barrels by her local garage introduced this material to her work. This made Sokari think of her environment and the products that are derived from oil/ petroleum. Every thing [sic] seems to have an element of this product, for example, on my desk, my keyboard, pen, sellotape, head phones, my jumper, and the concoction which has created my lipstick. ("About")

Douglas Camp's figures frequently wear bright lipstick themselves, the color popping against their metallic skins. In her sculptures *Open Blue Lips* and *Orange Lips* (both 2019), disembodied hands paint sets of disembodied lips, except the contents of the lipstick tube in each is a correspondingly colored toy car. The substitution makes explicit how the packaging of oil (as cosmetics, in your headphones, in your clothes) is a form of containment and concealment—a sly operation that inserts oil into "every thing." Our ready association of automobiles with oil (through gas stations, oil changes, and rubber tires) helps to make visible the less-obvious links between petroleum and its seemingly endless product list.

Elsewhere, Douglas Camp describes oil as a "magic product" ("Art Fights Oil"), for both its omnipresence and for its dangerous powers: "It's a magic product and yet it is killing us" (03:54). Reconfiguring oil barrels is one way to shake up our oblivious, ordinary relation to it and to reassert petroleum's place in the rank of "marvellous objects" (McDonald 353) where Roger Bacon placed it circa 1268: "yellow petroleum oil, that is, [oil] arising from a rock, burns whatever it touches, if it is prepared correctly" (McDonald 354). For Bacon and other medieval writers, oil's marvelous properties are not just medicinal or military, but the instigating fact of it magically "arising from a rock." When Agricola, the minerologist who is widely but wrongly credited with naming petroleum, *does* mention it, according to McDonald, it is actually in *De natura eorum quae effluunt ex terra* (*On the nature of those things which seep out of the ground*) (McDonald 363). This intrinsic quality of oil—the very nature of it being a thing that seeps—also appears in Douglas Camp's art, especially in her wrestling series from 2018. In each of these three sculptures, we see two figures, made, as always, of steel, wrestling on top of various oil barrels, one painted with both the Shell and the BP logos, another a mash-up of various oil cans (cooking oil, Gulf-brand oil), which reminds us of the many different kinds of oil that trail social and environmental spoilage behind them, especially in Nigeria.[8] All the barrels are leaking—or at least, they appear to be. Contrary to oil's true nature, the artificial seepage, also constructed from steel and found objects, is arrested.

---

[8] For more, see Yvonne Reddick's "Palm Oil and Crude Oil."

Progres

The very first word in Ken Saro-Wiwa's 1986 short story collection *Forest of Flowers* is Progres [sic], the name of the bus the young female narrator rides back to her small village of Dukana from college: "'Progres' was Dukana's pride, its only fast link with the modern world of the brick town where ships berthed and foreign goods were bought and sold. It made the journey daily and was much valued by all. It was proud witness to the progressive and co-operative, modern spirit of Dukana" (1). Misspelt, and thus undermining the claim to progress, the name inadvertently reveals the truth of Dukana as we come to know it—that it is in fact not progressive but backward and steeped in superstitious tradition. Although Saro-Wiwa begins his collection of stories with the image of a clearly focused and independent Nigerian woman on the move, the promise of Progres/s is not delivered on, even in the shift between part one of the book, "Home, Sweet Home," to part two, "High Life," ostensibly a relocation from the bare life (Agamben) of the Niger Delta to the modern metropolis. Recently, infrastructure theory has recognized this failure to progress as "stuckness": "the narrative sign that putatively universal terms like 'progress' or 'development' or 'place' or even 'climate' are yoked to a sense of collective pasts and futures unavailable and even unbelievable to people whose daily lives alienate them from any sense of a shared world, a shared future" (Foote). Variations on the theme of stuckness appear throughout petroforms in literal scenarios (feet stuck in the mud; cars stuck in traffic) that stand in for more abstract impasses. What Saro-Wiwa captures so well is the aggravated irony that not only are the bus and the road not symbols of progress, they compound the stuckness of those caught up in petroculture's mobility narratives, per Simpson and Szeman's definition: "Impasse in this sense names a continuation of the [situation as it is] wherein the overcoming of blockages cannot solve—and may in fact compound—the abiding stuckness" (80). The saga of Sokari Douglas Camp's full-scale bus replica corroborates Saro-Wiwa's sense of Nigerian progress as an oxymoron.

The public art project *Battle Bus: Living Memorial for Ken Saro-Wiwa* (2006) marked the twentieth anniversary of Saro-Wiwa's murder by the state. Douglas Camp's intention to mobilize his memory was stymied by authorities, who seized it at the port of Lagos in 2015 and have kept it impounded ever since. The bus was first on display in the United Kingdom for nine years and then shipped to Nigeria by Platform, a collective campaigning charity based in London. It was reported at the time that customs seized it on the grounds that it had "political value" and stated that the "inscription on [the] memorial bus is a threat to national peace" (Bassey). "I accuse the oil companies of

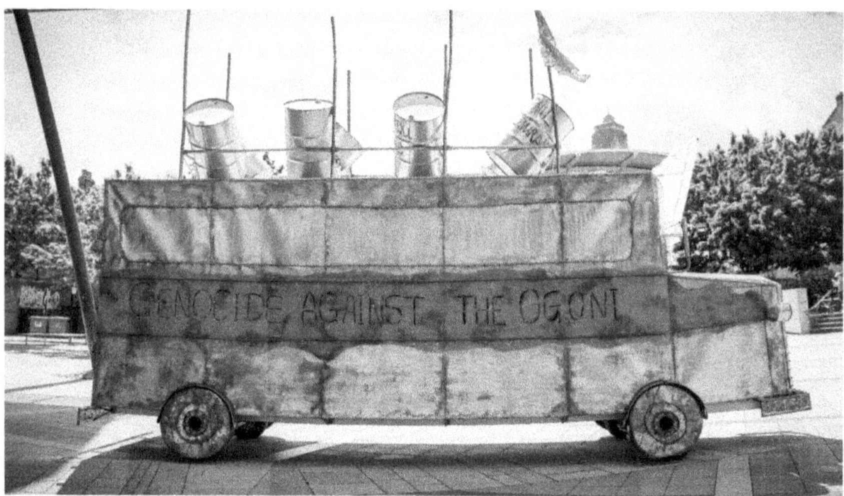

Figure 8. Sokari Douglas Camp, *Battle Bus* (2006), "a large-scale mobile interactive steel sculpture." © 2024 Artists Rights Society (ARS), New York / DACS, London.

genocide against the Ogoni," reads that inscription,[9] in giant, capital letters that perforate the large steel plates that comprise the roughly hewn outline of the bus, complete with steel windows and steel wheels. On top of the bus, eight oil barrels inscribed with the names of the other members of the assassinated Ogoni Nine balance akimbo. It is not a subtle piece. Rather, it is consequential, spatially and semantically.

From the original act of his murder, authorities have wanted to erase Saro-Wiwa's person, name, and legacy. His son, Ken Wiwa, in his memoir *In the Shadow of a Saint*, recalls his father's body's desecration by Sani Abacha, the military dictator who ordered his death:

> I'm not sure where or how the rumors that acid was poured on his body came about, but if it is true (and at the time of writing, this still hasn't been confirmed or denied) I imagine that Abacha wanted to wipe him off the face of the earth, obliterate him from memory. His paranoia about my father

---

[9] Although most sources, including Christiane Schlote in her essay on the bus, "Oil, Masquerades, and Memory," attribute this to Saro-Wiwa, Douglas Camp confirms that it is not a verbatim quotation but rather in the spirit of what he said many times, including in the ITV interview that inspired this version of it (personal interview).

was so intense that he was determined to prevent any kind of memorial being erected to him. (Wiwa 143)

Considering Douglas Camp's bus in the context of authority's demonstrated anxiety about memorializing and memory, it becomes clear that the primary threat the bus poses to the Nigerian state is remembrance. When the sculptor heard an interview with Saro-Wiwa in which he called what was happening in the Niger Delta "genocide," she says it "stopped me in my tracks and what I wanted was to stop people in their tracks" (personal interview). The state well knows that to stop the bus from moving altogether will also stop its ability to stop people in their tracks.

The politics of memorialization have never been more heated—consider the Rhodes Must Fall movement that began in 2015 in Cape Town, or the recent announcement that the Theodore Roosevelt statue that stood at the entrance to the American Museum of Natural History for ages (and about which Donna Haraway famously wrote her influential "Teddy Bear Patriarchy" essay) has been "recontextualized" in South Dakota (Schuessler). Those erections to the dead, white men of history were intended to be permanent and immovable, whereas with the Saro-Wiwa tribute, "the idea of a travelling memorial was conceived as an antidote to the colonial notion of fixed, figurative monuments," according to Dan Gretton, former Platform director (Arendt). Because a memorial by definition preserves memory, a mobile memorial complicates the objective. In *Memorials as Spaces of Engagement*, Quentin Stevens and Karen A. Franck define mobile memorials as "permanent in their materiality but temporary in their location" (60). On "making memorial objects mobile," they say, "this mobility opens up different relations between historical events, the passage of time, a commemorative object, meaningful relation to sites, and the localized engagement of various publics" (58). Giving examples like the AIDS quilt, they say, "In each case, the sudden appearance of a memorial at a new site gives it a degree of immediacy and novelty that a stationary memory can only achieve when it is first erected, and so the novel memorials are more likely to capture attention" (Stevens and Franck 59). This was certainly the case for the *Battle Bus*, which, according to Christiane Schlote, caused a stir when it was parked in front of Shell's London headquarters in October 2007 (251). In the UK, *Battle Bus* toured Bristol, Hull, Liverpool, and Birmingham; places chosen in part because of their historical links with the slave trade (Arendt). By dint of being moved through and between these locations, the piece sutures them into a quilt of "mourning sites" (J. Young, *Texture* 3) that covers also Ogoniland, the bus's intended last stop in Nigeria's Rivers State. In this way, not only does the "Living

Memorial . . . inform, agitate and inspire, and claim its space in the landscape of multicultural Britain" (Platform qtd Schlote 252), it also represents a post-colonial, global awareness of patterns of extraction, transport, and loss.

Because the bus is not actually a bus, but rather a "practical sculpture" (Douglas Camp qtd Schlote 257), it has increased capacity to contain symbolism beyond any normal use-value. Douglas Camp has commented in any number of interviews on the bus as symbol: "as a pollutant, a carrier of supplies and information, and as a metaphor for Saro-Wiwa's activism" (Arendt). All these concepts have to do with transport, which Douglas Camp singles out as an organizing idea both for the project and in environmental debate (Arendt). While the bus sets out to mobilize people politically, since it is not motorized, it cannot mobilize itself; it can only be moved by other means (it was transported around Britain by a haulage company). The double bind of being simultaneously mobile and immobile develops out of Douglas Camp's earliest work: "When I was desperate for an audience, it was when my children were quite small, and I put my work into a street where there was a local market. I dragged my work which happened to be a person with a pushchair and this thing that wasn't going anywhere but had wheels and had legs was in the marketplace seeming as if it was going somewhere" (personal interview). The artist is attuned to this and various other ironies attached to *Battle Bus*: "It is ironic that it is a fuel-guzzling object. . . . Can we have a lorry commemorating a campaigner against oil exploitation? Things are never black and white and intuitively I think Ken would like the irony of this because oil brought about his death but it is also playing a part in educating the world about the Ogoni people's plight and the other peoples of the Niger Delta" (qtd in Scholte). The bus's own plight is its darkest irony: being impounded by Nigerian customs means it has been arrested by law and arrested in movement.

The initial call for proposals from Platform described a competition to design a "living memorial" to Saro-Wiwa, where the language of vitality speaks to a desire to create "a unique piece of public art to keep alive the issues that Saro-Wiwa and the Ogoni fought and died for" and that the final product "not be a monument to a finished episode" ("Living Memorial"). This discourse, endowing the work with life, spills over into the reportage on what happened to *Battle Bus*, as we can see, for example, in an editorial entitled "A Living Memorial for Deadened Memories" by Nnimmo Bassey, an environmental activist who writes about its "kidnap" and "detention": "Seizing a sculpture gives the impression that the State is attempting to kill the message after annihilating the messengers." In that formulation, the Ogoni Nine were the messengers, and the inscription their message. If that message is alive—and can be killed—then it follows that the bus serves as its container for it, much as body does to being. It is the equivalent argument to John Milton's about books as living beings in *Areopagitica*: "For books are not absolutely dead things, but do contain a potency of life in them to be as active as that soul was whose progeny they are; nay, they do preserve as in a vial the purest efficacy and extraction of that living intellect that bred them." Just as a book is a vehicle of the life and spirit of its author, so too *Battle Bus* contains the efficacy and extraction of a living intellect.

"As in a Vial"

Before Douglas Camp used oil barrels in her art, she contained oil (and sometimes other substances that substituted for oil) "as in a vial" in various pieces, including *Rumble in the Jungle* (1998), *Assessment* (1998), *Close to My Heart* (1998), *Self* (1998), and *Guns and Oil* (1998). Schlote counts these as among Douglas Camp's "oil masquerades" (255), although they are not explicitly named as such the way some of her other sculptures are. In the first four of these pieces, a large steel figure holds up a rectangular glass box containing an image and oil. In the fifth, the glass container sits on a pedestal with two guns fashioned out of steel and wood crossed in front of it. The image encased in *Rumble* is a rural landscape of tall palm trees echoed in and extended outside the box by the miniature steel trees the figure also holds. *Assessment*, *Close*, and *Self* all use the same image, a photograph of a crude oil facility and nearby gas flare. Oil pools in the bottom of each of the containers.

Douglas Camp explains how the photograph came about:

> I'd been to Nigeria and taken a picture of a flow station and the people at the flow station in the creeks in the Delta rushed out because they could see we were filming them and had to tell us "No, no don't do that," and we just sped off on a speedboat, so it was quite a dramatic photograph. So, I brought this photograph back and I put it on acetate behind glass so it seemed very like the environment where this thing had been photographed and to put oil inside a container seemed like the perfect thing to do. (personal interview)

As explicated by the simile, the glass container stands in for the environment that the dramatic event, the flow station, and the oil all occupy. In their new setting, however, these things are forever frozen and can be replicated and manipulated into serving other purposes, such as art and activism. Putting oil into a container seems "like the perfect thing to do" because oil demands

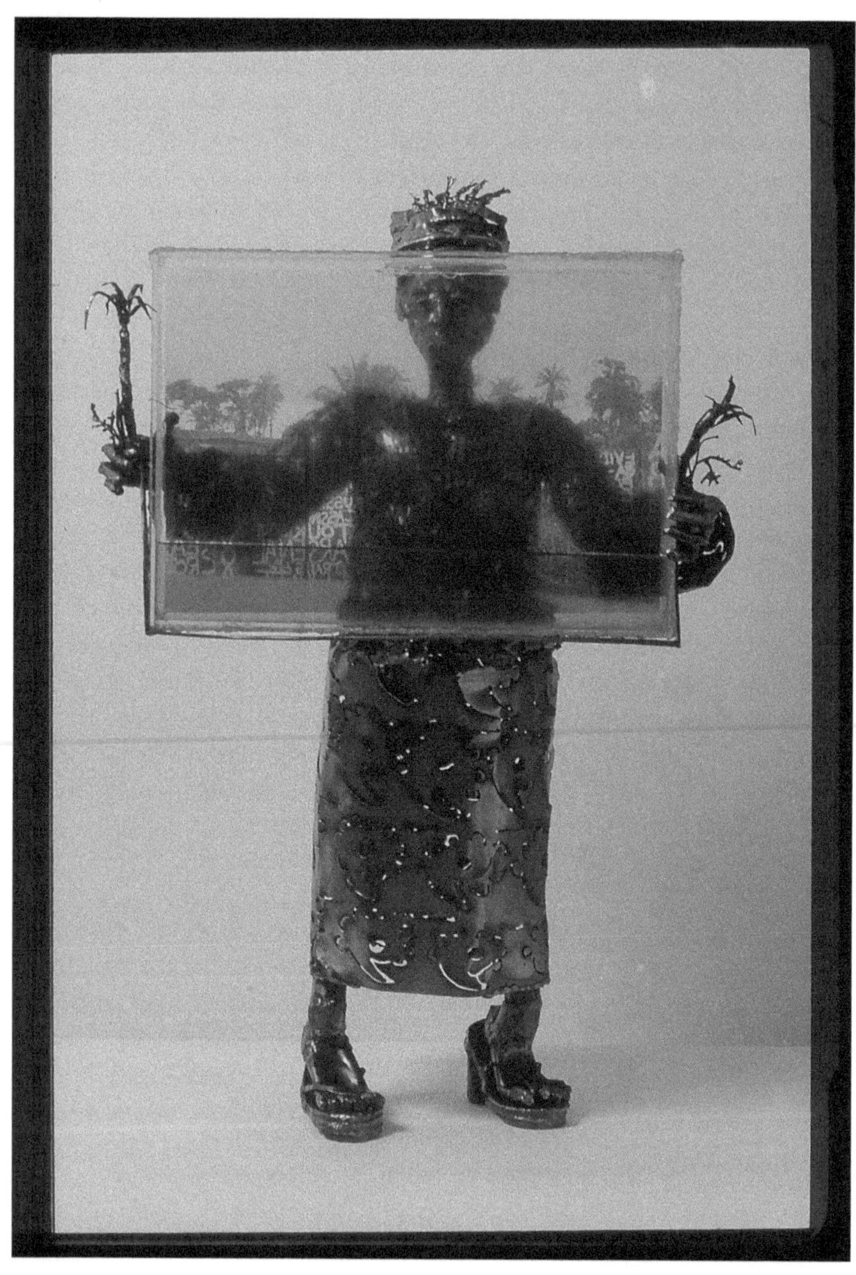

Figure 9. The trees the figure holds extend the landscape beyond the reliquary containing oil in Sokari Douglas Camp, *Rumble in the Jungle* (1998). © 2024 Artists Rights Society (ARS), New York / DACS, London.

a container. Containing it, per Milton, preserves "as in a vial" that which has been extracted—here, from the landscape—and which would otherwise seep out. Like in Botticelli's *Primavera*, where Douglas Camp sees "the landscape breaking up because something is so beautiful," she conceives of this practice as a confluence of beauty and violence: "A bit like stained glass, except that it was fire that I was putting behind glass" (personal interview). The containment of both the fire and what it represents (the dramatic act of escape, the spectacular flare, the environmental incineration) controls its volatility, tamping it down and subordinating it to the service of the sculpture.

In *My World Your World* (1997), another steel figure holds the same photograph, this time sans oil or glass case. Douglas Camp gives the backstory:

> The National Museum of Ethnology in Osaka, Japan, . . . wanted a work that particularly represented [the] Kalabari people. We the Kalabari, carry photographs of a relative after they have been buried to let people know that we are related to the deceased and they have passed away. After being in Nigeria for my holidays I could not help but not notice the increase in the number of oil wells in the Kalabari region. An oil well had been sunk near an island that is particularly sacred because this was where the Kalabari nation started. From this island Kalabari chiefs spread out across the delta with their people to inhabit the islands we now live on. It was always nice to go to this island and to walk around the palm groves and the mounds where our people are buried. Now we have a roaring oil well beside the island and so much pollution. . . . So I thought I would work with the reality of my home surroundings, a picture of our environment. Instead of a picture of a relative I have an oil well, which a female figure holds as if she is holding the image of a loved one. (*My World Your World*)

What has passed away is a landscape and a lifestyle, but the photographic relic, normally a tangible reminder of what has been lost, pictures instead what has replaced it—the "reality" of an environment filled with roaring oil wells and pluming pollution. In a twist, then, Douglas Camp gives us a kind of living reliquary memorializing what persists and haunted by what it has killed off.

Holding a picture of an oil well "as if . . . of a loved one" dramatically recalibrates the relationship between person and system, closing the distance that the corporate world would prefer to maintain so as, among other things, to be able to claim plausible deniability for the immediate and long-term damages it causes. Denying distanciation establishes intimacy—in this case the specific infrastructural intimacy of living in, among, and with the energy infrastructure itself. In conversation, Douglas Camp speaks of the corporation as relative: "Shell is at the forefront of Nigerian employment as well our destruction. It's quite a relief to talk about them because they are a naughty relative that's upset people. It's not people from another planet" (personal interview). Hugging the corporation close, like a naughty relative, means they can't get away; they can't get away with it.

The discursive move of treating the nonhuman as relative recurs in her work and serves as a way of reasserting the value of human life in the face of dehumanizing narratives. Of a 2010 series inspired by BP's Deepwater Horizon oil spill in Florida, which includes the sculpture *Relative* (in which two human figures hold a gold-framed image of a pelican), Douglas Camp acerbically observes: "These pelicans were on all the front pages in the world, and I thought, well, Kalabari people have never been seen, but these ugly birds are everywhere—so I made these pelican portraits, saying, oh, these are my relatives" (personal interview).[10] Combined, these relational gestures have less to do with going against the grain of the recent nonhuman turn in academia, the law, and elsewhere (which would attribute agency and confer rights to fauna, flora, and landscape features[11]), and more to do with redressing a status quo that demotes the human by giving preferential treatment to the other. While making visible the human experience does pack an implicit protest against those who would prioritize the nonhuman, whether out of avarice (sinking new oil wells) or altruism (saving oil-soaked pelicans), it does not foreclose kinship-making with other species. On the contrary: Pelicans in particular represent a special interspecies entanglement. Although we need to go back 320 million years to find a common genetic ancestor, we can literally relate ("Amniotes"). Plus, the bird carries enormous symbolic attachment for us, signifying Christ-like mutual aid and sacrifice, as popularized in Botticelli's age in sentiments about "the pelicane who striketh blood out of its owne bodye to do others good" (Saunders). Although Douglas Camp did not have pelicans in mind when she said this, this is "what this work is all about: hidden connections, ties, links and roots" (Houghton, "FW" 4).

When Douglas Camp reminds us that the people who work at Shell are not "from another planet," she reasserts their humanness, if not their humanity. Neither alien nor abstract, they are real people, making real decisions, with real impact on real lives. Although it is true that "the transnational regime

---

[10] Per the BBC article, "The Oil-Soaked Bird That Shocked the World": "Now, when Americans think of the BP Deepwater Horizon oil spill, they likely envision an oil-soaked pelican."

[11] See, for instance, Nicole Rogers on "wilding."

of petrochemical extraction and petroagriculture is the chief engine of the Anthropocene," seeing human relations through the lens of a "planetary pursuit of pure power and pure profit" risks alienating us from lived experience (Mbembe, *Night* 44, 46). In contrast, a work like *My World Your World* brings us back to earth, where the local and the global run together (aided by the lack of a customary comma in the title). Because we understand that the figure holds an image of the deceased, it announces the death of a blighted landscape, either in despair because oil wells and their fallout have killed things off, or in anticipation that eventually the oil industry will be done away with. Putting the scene in the hands of the figure reduces its relative importance (so to speak) so that the site is held within that relationship, doubly contained by the borders of the image and the superseding embrace of the human. Because she is a female figure, displaying the often-hidden gendered labor of familial relations and their remembrance, she also represents another mode of petrofeminism.

The human figure, whose steel body is simultaneously surface and container, takes pride of place. With steel as the medium, each figure is a double container—oil drum, barrel, or can turned into bodily form. Each container is also its own surface, the taut limit at which the gaze stops, seeing race, notwithstanding the deceptive neutrality of the metallic finish: "Steel is also nearer to being Black. I like the colour of it. I like how I can polish it up and the darkness of it. Clay or marble don't appeal to me in the same way" (Douglas Camp qtd James). Color races steel but so does its object biography, as the matter of "extraction regimes" (Yusoff), as the building stuff of empire, and as the container for the dissemination of petrofluids, all of which coalesce in a history and experience of Black materiality embodied by "that metal-man, that silver-man, that wood-man, and that liquid-man, that body of extraction destined for endless transfiguration. A vessel both metaphorical and plastic" (Mbembe, *Necropolitics* 164).

"Water No Get Enemy"

The Akeley Hall of African Mammals at the American Museum of Natural History (AMNH) is full of sculptures, of a kind. Carl Akeley's original dream of being a sculptor took shape as his practice of taxidermy, both arts "narrative tools" that helped him tell "some of his stories in bronzes as well as in dioramas" (Haraway 37). The hall consists of twenty-eight habitat dioramas "based on the meticulous observations of scientists in the field in the early 20th century and the on-site sketches and photographs of the artists who accompanied them. They feature animals set in a specific location, cast in the light of a particular time of day" and surround "a freestanding group of eight elephants, poised as if to charge" ("Akeley Hall").[12] In a specific location, at a particular time of day, poised on the brink of action—the diorama suspends all space-time coordinates in the aspic of the natural history museum.

In 1998, when Sokari Douglas Camp's masquerade sculptures were exhibited at the AMNH alongside Yoruba and Kalabari masks and headdresses from the museum's holdings, one critic worried that "this juxtaposition potentially invites the idea that Camp's [sic] work should be separated from the realm of fine art and relegated to the role of anthropological object. Furthermore, her work is on view in a museum better known for its dioramas in the Man in Africa Hall than its forays into contemporary art" (Barnwell 81). One might ring a similar alarm bell about her inclusion in the ethnological museum in Osaka. In the end, though, the critic concluded that "Camp's [sic] sculptures should be viewed as art rather than mannequins of actual performers" because of how she "appropriate[s] forms, creates new structures, and reinvents meaning" (Barnwell 82). Borrowing the name "masquerade" from dynamic performance to use for still sculptures captures this counterintuitive crossover of forms.

Water No Get Enemy: Counter-Cartographies of Diaspora,[13] a short "film essay" by Remi Kuforiji, opens with a mythological reference[14] to "an ancient story of a figure, a woman, a deity" (00:03) who watches men masquerade and then enables a segue to the film's present day of 2021, in which men perform "a new masquerade: a method of cartography that critiques harmful extractive practices by bringing multiple diasporic sites into dialogue through performance," as described by the film's webpage in the Disembodied Territories site (*Water No Get Enemy*). The film features a cameo of Douglas Camp, who returns to her theme of outside, alien influence, asking: "You do realize that masquerades are like visitors from outer space? It's not even that this thing is from inside the Delta; it's something that was a visitor" (02:35). She approaches "this thing," the masquerade tradition, as fluid and plastic, available for modification as it

---

[12] This arrangement has not changed since Haraway wrote about it in 1984, increasing our perception of the permanence of these displays and their "natural" histories.

[13] The film borrows the title *Water No Get Enemy* from a Yoruba proverb whose literal message is that water is essential and only a fool would be its enemy. More figuratively, it teaches that if we live in harmony with nature our future will be assured ("Water"). It is also the title of a Fela Kuti song from 1975.

[14] The oft-repeated origin story for the masquerade tradition has it that, "According to myth, Kalabari women were the first to witness the water spirits playing by the river on the fringes of the mangroves. They told the men, who later enacted the performances" (Barnwell 81). This is as told in Robin Horton's "The Kalabari 'Ekine' Society." Horton, a British anthropologist, was Douglas Camp's guardian when she was a minor.

has always been: "As masquerades are all about men impersonating both male and female 'characters' gender becomes quite a fluid concept, and my own masquerade-inspired sculptures have always exhibited this duality" (Douglas Camp qtd Houghton, "FW" 4). Whereas Douglas Camp's masquerades mix gender as well as genre, the film reiterates traditional gender roles, despite, or perhaps because of, a desire to learn from "indigenous knowledge as a serious form of spatial practice" (*Water No Get Enemy*). Kuforiji, when asked about this during a webinar, said that he was trying to be sensitive to the legend in which women are the instigators and chief audience of masquerade. If the audience identifies with that viewing position, its gender roles will also be in flux.

These new masquerades—Kuforiji's film and Douglas Camp's masquerade sculptures—share an interest in teasing the relationship between stillness and movement, like the diorama, but their politics differ drastically. The preservational politics of the diorama and its craft of taxidermy look backward to "remembering the perfect experience" (Haraway 36), the apex of the hunting safari. The masquerades look back instead to make connections across time and space, to reveal that the site of extraction and its extractors are the same (Kuforiji, webinar). To establish this, they use "ubiquitous things"—the commodities of oil, plastic, cotton, and steel as a common language (Kuforiji, webinar). Douglas Camp hopes her masquerades will "give the viewer a frame for, or a picture of, what a Kalabari performance is like" (*Big Alagba & Sekibo*). But the piece she describes—two figures side-by-side—has no frame. And Kuforiji's film-essay-ritual-theater is actually a diptych (although we cannot see that as shown online), doubling up on itself so that it also breaks free of a single frame or form. Self-aware, these artists know that they cannot make containers that can hold the fluids of the Niger Delta, "now dominated by another materiality, a viscous film of technicolor indeterminacy—crude oil" (Kuforiji, dir. *Water No Get Enemy* 02:48). They know that, unlike the diorama, they cannot freeze a premonition of movement (Haraway 23) that has already happened, and is underway.

CHAPTER 4

———

PETROHORROR

———

"A Story That Evokes Horror"

In Michael Watts's comprehensive account of Nigeria as a petrostate, "Sweet and Sour," he suggests that "the complexity, diversity and magnificence of the Niger delta is best appreciated from the air." Zooming in on its oilfields, however, shows "a bleak picture, a dark tale of neglect and unremitting misery." Illustrating this dark tale, Watts goes on: "One of the horrors of the delta is that the ultra-modernity of oil sits cheek by jowl with the most unimaginable poverty. Around the massive Escravos oil installation with its barbed wire fences, its security forces, and its comfortable houses are nestled shacks, broken down canoes and children who will be lucky to reach adulthood" (Watts, "Sweet and Sour" 40, 37, 44). In the documentary film *Daughters of the Niger Delta*, one of the participants, Hannah, speaks of poverty as a presence, like a monster looming over her: "What I want to change now from my life is poverty. I don't want poverty to come near me. Because poverty is dangerous. It can lead you to something that you don't want to enter inside" (05:36). Poverty personified approaches, attacks—like something from a horror movie that might catch you and lead you astray, deeper into something unnamed and unknown, farther away from safety.

This special kind of horror, emerging in the gap between the satellite view and the "reality on the ground" (Watts, "Sweet and Sour" 39) and between the extremes of the oil experience, is what Ralph Nader means by "petrohorror," a coinage he used when endorsing the 2009 documentary film *Sweet Crude*: "Add the petrohorrors in the Niger Delta to the 'price of oil.' There is nothing 'sweet' there, but the oil industry's profits" ("Sweet Crude Buzz"). There is a small but significant number of documentaries about the environmental degradation, cost to human life and livelihood, and accelerating violence in that oil-producing region of Nigeria, and they merit being considered together not just as documentary films that show this horror as content, but as a cinematic form we might call *petrohorror*.

The films read here share a sense of purpose with other modern documentaries and other

> films about oil [that] understand themselves as important forms of political pedagogy that not only shape audience understanding of the issues in question, but also hope to generate political and ecological responses that otherwise would not occur. This production of an outcome or change in societal imperatives is a long-standing desire of the kind of politically and ethically committed documentary filmmaking that for publics has to a large degree become identified with the function of documentary as such. (Szeman 424)

But the oil documentary's lineage traces back to quite different origins, where the function of the documentary was not that of political pedagogy but a strategy of corporate persuasion. This history comes to us from the "Sponsorship" section of Patrick Russell and James Pier Taylor's *Shadows of Progress: Documentary Film in Post-War Britain*, which is then globally elaborated on in Mona Damluji's chapter in *Subterranean Estates: Life Worlds of Oil and Gas*. What these accounts make clear is that the galling corporate cynicism apparent today in pro-petroleum advertising—like greenwashing campaigns—was present from the start; oil companies' earliest petrofilms (Damluji 149) were produced to promote an image of the corporation as a "benevolent influence" (Canjels qtd Damluji 154) through ostensibly honest, factual, and educational representations. Damluji describes the ways in which, by erecting a faux screen between the product itself and its uses, applications, and infiltration in the world, Shell in particular (but other petroleum companies too), was able to create and then instrumentalize the entirely new industry of film to extend its influence over public perceptions of the oil industry.

"The history of cinema is a history of oil," writes Damluji. This linkage extends from product to production to passage to production: "Petroleum was integral to the development of early photography and the creation of celluloid filmstrips. Beyond petrochemicals, oil as fuel was essential for transporting film pioneers and their heavy equipment to remote locations around the world" (Damluji 147). Shell's film unit in the early twentieth century ushers in "the first genuine international documentary film movement" (Elton qtd Damluji 155). Although Damluji warns against conflating the two competing varieties of oil documentary—the industry's "petrofilms" and the politically and ethically driven documentaries I study here—we can see in both the "constitutive ways that the politics of petroleum have shaped the cultural production of documentary images of modernity" (Damluji 148).

The premise of *Petroforms* is that oil contorts and distorts existing aesthetic forms; in the case of the documentary film, oil shaped the form from the outset, but then, as we shall see, even that autogenous form cannot accommodate the surpluses of petroleum production and problematics. These films intend to overturn the petroleum-celluloid conspiracy that defines their origin stories, yet they are still indebted to the form and its formation.

Ordinary Horror

Seeing oil as fundamentally constitutive allows us to name it in all its manifestations. It is probably not coincidental that two other recent book chapters also employ the term *petrohorror*, as we collectively label more and more aspects of modernity with the petro- prefix. My theory of documentary film as petrohorror, though, reads against the grain of the others. In "An Oily Mirror: 1950s Orang Minyak Films as Singaporean Petrohorror," Yogesh Tulsi argues that those popular films are indicative of the moment when fossil fuels are recognized as something monstrous. The "orang minyak," a shiny, greasy creature haunting the *kampung*, or village, is a monstrous representation of modernity's ills. In "Fade to Crude: Petro-Horror and Kubrick's *The Shining*," Pansy Duncan reads for petroleum's status as the horrific object of *The Shining*'s central scenes of "gross display" (Linda Williams qtd Duncan 186). Both offer useful insights, but whereas they focus on the spectacle of horror, I propose that what's really scary is the ordinary horror of the oil experience as captured in the brute reality of the documentary film.

Joshua Carswell, in his blog *orbistertius*, talks about the role of oil in genre horror literature, usefully pointing out that Amitav Ghosh, in his field-defining essay "Petrofiction," says that the story of the oil encounter "is a story that evokes horror," among other reactions. Carswell builds on Graeme Macdonald's long essay on depictions of energy resources in science fiction, which argues that images of environmental degradation "compose a horrific form of petro-sublime," which is insufficient to register the enormity of the problem: "Only by considering the horror of oil's banality can we begin to register its everyday use as a central substance in the environmental fantasy that is late capitalism" (Macdonald, "Improbability Drives"). The modes of horror that relate to our energy crisis and petroculture are not only those of the supernatural Singaporean "oily man," or the sprays of blood across the Kubrick camera lens, but those that register the horror of its banality.

This discussion concentrates on four films, all made around the same time with common themes and purposes, but which can be classified differently by a taxonomy of direction and production. Two, *Sweet Crude* (Sandy Cioffi, 2009) and *Delta Boys* (Andrew Berends, 2012), are by American filmmakers; two, *The Naked Option* (Candace Schermerhorn, 2011) and *Daughters of the Niger Delta* (Ilse van Lamoen, 2012), are made in collaboration with Nigerian filmmakers.[1] They all present as conventional documentary films, but when we look more closely, we can see those conventions as aspects of horror: In addition to the monsters of oil and poverty, there is the eco-horror of unnatural environments, the body horror of physical degradation, and the psychological horror of uncanny calm in tension with outright violence.

Despite sharing horror's hallmarks, the films take radically different approaches to the problem of representing petrohorrors on screen. In *Sweet Crude* and *Delta Boys*, the two Global Northern films, it is almost as though the documentary filmmakers consulted one of the checklists of horror tropes that abound online, and in doing so, lent a familiar form and structure to the otherwise overwhelming, all-encompassing, structureless condition of living with and in oil. In contrast, *The Naked Option* and *Daughters of the Niger Delta* clearly hope to subvert the documentary-as-horror model, though oil's banality inevitably compounds the horror.

Whether or not they deploy conventional cinematic horror tropes, corporeal and psychological horror constitute the worlds of all four films, often in the form of specifically gendered violence. The gender discrepancies that form horror's backbone clichés pertain equally in the Delta swamps. We find ourselves in a realm where anyone, especially a girl or a daughter, might fall victim to what awaits; "It was as if as a female child, anything can happen to you" (*Daughters* 23:50). But mitigating those horrors are an arsenal of petrofeminist strategies and practices, from the sweet tone of *Sweet Crude* to the deployment of nudity and women's bodies in *The Naked Option* to protest the exploitative petromodernity that offers no benefits to them.

In Chapter 2, on romance novels, we saw a kind of petrofeminism that desires more career options for women, which goes hand in hand with women participating in expanding local roles in Nigeria's oil industry. To the extent that documentary oil exposés are an offshoot of the oil industry, the making

---

[1] The credits and poster for *The Naked Option* list Sam Olukoya as "Co-Producer Nigeria" (although his name does not appear in the IMDb). *Daughters of the Niger Delta* was a "bottom-up production," in which local women were taught filmmaking through a program called FEMSCRIPT, funded in part by the Catholic relief fund Cordaid, which works "in and on fragility" (Cordaid). These are complicated structures of financing and production, and they bear reflecting on when we think about the documentaries as collaborations between locals and outsiders.

of *Daughters of the Niger Delta* fits that model. *Daughters* was made by local women who had never made a film before, trained via a participatory video project that ran for one year in the Delta. It is an exposé of the taken-for-granted horrors that Delta women endure, from mud and malnutrition to mortality. At the start of *Daughters*, viewers are told that, "Our story is different from the usual media reports about oil outputs, conflict, and kidnapping. It focuses on the everyday lives of women" (00:26). About halfway through the film, the voiceover continues:

> Every year, one out of every seven children in the core Niger Delta states is estimated to die before the age of five. The number of under-five deaths is over 65,000 per year, out of a total population of almost 12 million people. These deaths far outnumber the casualties of the much-reported violence and kidnappings in the region. Yet, we hardly ever hear about them in the news. (27:28)

The documentary's antisensationalist approach reports reality matter-of-factly, but the statistical ordinariness of the horror makes it that much more horrible.

Whereas Singaporean oil monsters figuratively embody fossil fuels, and we have to trace the manufacturing history of fake blood and lipstick back to petrochemicals in order to arrive at the claim that oil is the true focus of *The Shining*, no such obfuscation happens in the oil documentary, where oil is everywhere, excessive, and explicit. It does not need to be allegorical, or coded through other commodities. It is present in the ooze between the bare toes and flipflops of the Nigerian villagers, gingerly balancing on boards to walk between huts in a muddy swamp; it is present in the slicks on the river as the fish float up to the surface, dead; it is present in the toxic flares in the background of every shot; and it is on the surface of the bodies that are slick with it and sick with it.

Uncanny Calm

The first words director Sandy Cioffi says in her voiceover in *Sweet Crude* are "This is not the movie I intended to make" (00:03:00). What she means is that she set out to make a movie about a library being built in an impoverished Niger Delta village and ended up making something quite different. The *Variety* review of *Sweet Crude* emphasizes its unusual qualities, including "Cioffi's frighteningly gentle narration," as its most marketable ones, "all of which blend to seductive psychological effect, suggesting gossamer dreams about paradise lost, with an undertone of unrefined fury" (Anderson). The narration frightens because its gentleness contrasts with the inhumane subject matter and because it takes advantage of classic horror movie aesthetics, where sound manipulation, including "uncertainty, expectations (silence during tense moments), whispered voices, [and] context disconnect," is used "to evoke physiological responses" (Michael Epstein qtd Fu 38). Whispers and other otherwise innocent sounds like children's laughter become terrifying in horror films, making us "feel discomforted" (Fu 40–41).

Sweet Crude opens with an assumption of untranslatability: a scene of women and children laughing and chatting that does not get a translation or subtitles (unlike other exchanges, later). The first English we hear at any length or legibility is an American National Public Radio broadcast discussing the "sabotage" campaigns of MEND (the Movement for the Emancipation of the Niger Delta, one of the area's largest and most organized militant groups) and their disruption to the international oil supply (*Sweet Crude* 00:02:20). While the NPR story doesn't represent the attitude of the filmmaker, it does give the first word to a general attitude of willful ignorance Cioffi has to fend off. In order to do so, she adopts an intentional posture of naivete—delivered in the uncanny calm of her tone of voice—to bring a genuinely naive (or else cynical) Global Northern audience along on the path of learning where their oil comes from and what is happening in the Delta.

With that as her starting point, the film is less a "document [of] life in the Delta" (*Sweet Crude* 00:11:10) and more a narrative of a political education. Certainly, none of the contextualizing historical timelines, backstories, or colonial framework would be necessary for a local or better-informed audience. This is not to say that documentaries shouldn't educate and inform, but rather to point out that, by channeling an attitude of wide-eyed incredulity, she sounds a bit like a parody of the "white saviors in Third World narratives," described by Nigerian-American writer Teju Cole, "who make the story more palatable to American viewers." In her director's statement in the film's press kit, Cioffi recounts: "I returned home and the more I thought about it, the more stunned I was: *How could this story be playing out in the most populous country in Africa, the world's seventh largest oil exporter and arguably a strategic lynchpin in the stability of all of Western Africa—and not make the front page of every international newspaper?*"

We know that her naivete is an assumed position, an angle Cioffi uses to reflect her likely Western audience back to themselves, because, for one, she is self-critical about the role of documentarian as savior. About eleven minutes in, she says, "I thought it would be irresponsible to assume that I knew why things have gotten this bad and arrogant to think my presence might not do more harm than good" (*Sweet Crude* 00:11:10). In addition, we see the risks that she and her project were subjected to in the making of the film. The epilogue informs us that, on returning to Nigeria from the United States in 2008 to finish it, Cioffi was detained by Nigerian authorities and reels were seized for flouting a ban on traveling through the "military zone"[2] of the Niger Delta without permission.

In a horror film, a character's soft tone might be designed to seem unassuming or meek but hide the menace within ("Soft-Spoken Sadist"). Here, the menace is twofold: from the crude oil and from a tempting redemption narrative in which the filmmaker saves the day, a narrative in which she wonders whether "telling this story" might actually impact the situation and "if this time, just maybe, an African tragedy could be averted" (Cioffi, Director's Statement). Cioffi's "frighteningly gentle" affect counters that narrative, suggesting, in its use of social constructs, a strategic, even gendered, approach to the very making of the documentary. She cedes authority over her film in more than one way, first by entering into a negotiated filmmaking process with her subjects, who ask, "How is the movie going to look like? Is it just an interview with the people alone?" (00:17:50). After hearing some requests, Cioffi changes the subject:

> CIOFFI: You said we should go out fishing . . . one day?
> FANTY: Yeah.
> CIOFFI: We should film that.
> FANTY: Yeah. Well, let's see how it goes. (00:18:30)

Eventually, Cioffi goes so far as to intercede on behalf of the local activists:

> Two days later we were back in Warri. Joel called a meeting to talk with the young men of MEND about the massacre and their threat to retaliate. They said they would stand down but only for the promise of peace talks monitored by a credible third party. We met over the next few days and drafted a document describing the issues and steps to resolution as they saw it. They asked us to deliver this document to all of the stakeholders in the region as well as the international community. And they asked us to bring media attention to the crisis. The movie I was making was changing again.

---

[2] In 2008–09, security forces declared much of the vast wetland region a military zone and barred outsiders from traveling there without express consent by authorities ("Four in Seattle-Based Film Group Arrested in Nigeria").

In some ways, it felt like it wouldn't be a movie at all anymore. Tompolo had
urged me to step in as an ambassador and MEND thought I could somehow
help obtain third-party intervention. I knew that taking on either of those
requests could mean stepping outside the acceptable role of a documentary
filmmaker. (*Sweet Crude* 00:48:00)

Having stepped outside of "the acceptable role of a documentary filmmaker,"
Cioffi feels she is not making "a movie at all anymore." By inserting herself into
the action, she checks herself as potential savior and marks the limits of what
we usually see documentary doing, rupturing its usual form.

Sweet Crude's artwork reminds us that the oil documentary is a byproduct of petromodernity. Paintings that emulate close-ups of rusted oil barrels
feature in the title sequence; in the historical timeline, where Africa blossoms
like a rust stain across the screen; and during the film's epilogue. In the end,
Cioffi leaves us not with the story of the Niger Delta (that story is on the tapes
confiscated by the authorities), but with her own. In the epilogue, about the
detention of Cioffi and her crew, a transparent overlay of stain or spill spreads
over the swamp scenes, gaining opacity until the mangrove trees aren't visible
anymore. The slow creep covers, obscures, mimicking the takeover of a presence in a horror film, and leaving no doubt that this is a movie about oil. But it
was always already an oil movie. The library was going to be funded by Chevron
(*Sweet Crude* 00:4:45).

Violence as Horror

In both *Sweet Crude* and *Delta Boys*, much of the interstitial footage consists
of the point of view from the front of a dugout canoe. From that viewing position, we get a virtual tour of mangrove swamps and delta tributaries, tranquil
scenes, but always with flares burning in the distance. The tourist perspective
of the films' visuals matches their epistemological perspectives. Like Cioffi,
Andrew Berends, maker of *Delta Boys*, "had just a basic understanding of the
nature of the conflict" (00:01:08) when he began his project.

Although the two films have many things in common (scenes of village
life interspersed with news footage, the American filmmakers' commentaries
alternating with local spokespeople, and so on), where Cioffi's film exploits
qualities of a conventionally female voice, Berends's is a celebration of militant masculinity. "I felt an attraction to these men who were taking on the
world's largest oil corporations and the corrupt government of Africa's most
populous nation, so I took my camera to Nigeria to try to meet them" (*Delta
Boys* 00:02:24), he tells us at the start of the project on the Delta militants.

That attraction manifests itself in shots that feast on displays of weaponry and a jumble of cultural references, from gas masks to balaclavas. One scene opens with a close-up of men cleaning and oiling their machine guns, upright between their legs. If not homoerotic, the moment is certainly homosocial, and is interrupted when one of the men directs the director: "Andy, ask me a question" (*Delta Boys* 00:15:36). As Berends asks him about himself, he continues to work on his weapon. The mood is intimate; the young man's voice sounds like it breaks when he shares that he's an electrical engineer by training, waiting for his government to give him a job while he bides his time in the camp (*Delta Boys* 00:16:14).

Having established the intimacy of the filmmaking relationship, Berends plays up the theme of a band of lost boys who have found a substitute family and a father figure in the vigilante group's leader. A long fight scene between two vigilantes ends with them standing before him, getting scolded. *Delta Boys* is by far the most beautifully shot documentary of the ones I look at here, full of glistening hard bodies, lovely lighting, and pretty patterns of oil on water, but that fight is one of multiple scenes of violence in the movie. The camera doesn't shy away from men getting whipped for dereliction of duty, but its shifting focus to contrast with the spooky normalcy of nearby fishermen makes the brutality all the more disturbing.

The whipping scenes are two of the more protracted sequences in the film. Why cut away from the gruesome cattle-killing scene but show those whippings in such extended, exquisite detail? Achille Mbembe's comments on slavery suggest an answer:

> The slave's labor is needed and used, so he is therefore kept alive, but in *a state of injury*, in a phantom-like world of horrors and intense cruelty and profanity. The violent tenor of the slave's life is manifested through the overseer's disposition to behave in a cruel and intemperate manner, as well as in the spectacle of pain inflicted on the slave's body. Violence, here, becomes an element in manners, like whipping, or taking the slave's life itself: an act of caprice and pure destruction aimed at instilling terror. (*Necropolitics* 75)

When we superimpose Mbembe's words onto the petrohorror documentary, his description tracks with what is shown on screen: The film itself is a phantom-like world of horrors, a phantasmagoria that suspends the militant in a state of injury, which we experience as a spectacle of pain. The violence of the whipping disturbs us for the same reasons violence disturbs in horror movies—when it is "shown or perceived to be frightening, disgusting, perhaps even nauseating, i.e., when it denotes brutality, sociopathy, sadism, and other socially undesirable traits" ("Violence"). Despite Berends's best intentions ("to help raise awareness of the Niger Delta oil conflict" [Berends]),[3] it is hard to ignore the multiple ways in which the Black male body is being violently exploited: as object of the film's homoerotic gaze, as slave to the rebels and their overseer, and as pawn in the geopolitics that sanction militancy as one release valve for the Niger Delta's discontent. Like a perversion of petrofeminism, this version of petromasculinity[4] also embraces the romance of petromodernity, but its care economy takes place in a rebel camp and its solidarities are forged through displays of hypermasculinity and ultraviolence.

Body Horror

Even when Berends isn't showcasing hypermasculinity, his film frames events in terms of male experience, including during a childbirth, the second main scene of the film. The birth itself is not shown; rather, we see a frantic flurry of activity as villagers try to save the mother and child during a risky delivery, during which the camera's attention is largely given over to the laboring woman's father, impatient for baby to arrive so he can pull in his fishing hooks. Even void of the actual birth, the scene is replete with various iterations of body horror.

Body horror contains multitudes, including the subcategory of "womb horror," one version of which is the "imperiled in pregnancy" trope which, while "especially shocking for most viewers/readers [o]ften turns out all right in the end, as both audiences and creators frequently find it unpalatable to play this trope through to the end" ("Imperiled"). As the panicking villagers scoop sand up with their hands nine minutes into *Delta Boys*, we can only imagine what home remedy they are attempting behind the curtain, and the imagining makes it worse. The expectant mother, Mama, whose name identifies her entirely with her maternal body, has a skin condition that raises thoughts of contagion and contamination. The eruptions on the surface of her skin complement the deep rupture happening within her body, inside the hut. It doesn't

---

[3] In March of 2019, Berends committed suicide after a Parkinson's diagnosis. One tribute to him, which said, "He lost himself in conveying the stark realities of other people, in becoming their voice, their witness," speaks to the horror of giving voice to horror (Longley).

[4] In distinction to Cara Daggett's definition of petromasculinity as "the hypermasculine mode of support for fossil fuels in the rising authoritarian movements in the U.S. and elsewhere" ("Petro-Masculinity and the Politics of Climate Refusal").

matter that the childbirth is off camera; written on her face in pustules and pain, it disrupts the inside/out boundary of the abject maternal body in true horror fashion (Aldana Reyes 75). It does turn out fine in the end, at least for now, but at the end of the film, we see Mama again. She's pregnant with a second child, the process threatening, horrifyingly, to repeat itself.

Nothing Is Scarier

In horror, nothing is often scarier than something. Conventionally, this is a "horror trope where fear isn't induced by a traumatic visual element or by a physical threat, but by the *sole lack of event*" ("Nothing"). In Candace Schermerhorn's *The Naked Option: A Last Resort*, women wearing nothing are the source of horror for the male oil workers they confront. *The Naked Option* documents the "last resort" efforts made by Delta women to intervene in the oil politics of the region by stripping naked to call attention to their demands in their occupation of a Chevron facility in 2002. Standing before the person who wronged you naked is a curse, according to the film; "the threat of nudity" is such a taboo that "that man should consider himself as almost dead. It was like handing the person over to the demons" (*Naked Option* 00:35:22), explains Felicia Nwlutu. Nwlutu's simile becomes another expert's metaphor: "Women when they organize and kill you don't see any blood but the person is dead. It's a metaphor" (*Naked Option* 00:34:52). Whether purgatory or death is figurative or literal, the naked protesters invert the power of the taboo, taking the

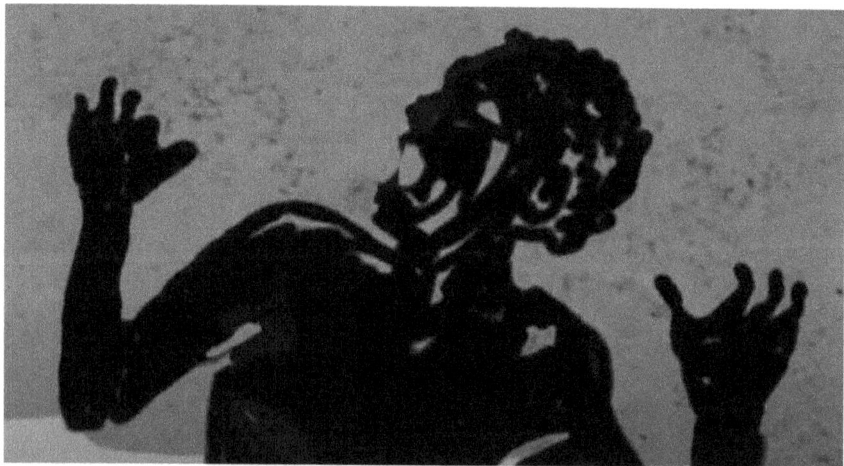

Figure 10. Facing the cultural threat of nudity. Artwork from *The Naked Option*, Deja Hsu. Dir. Candace Schermerhorn, 2011.

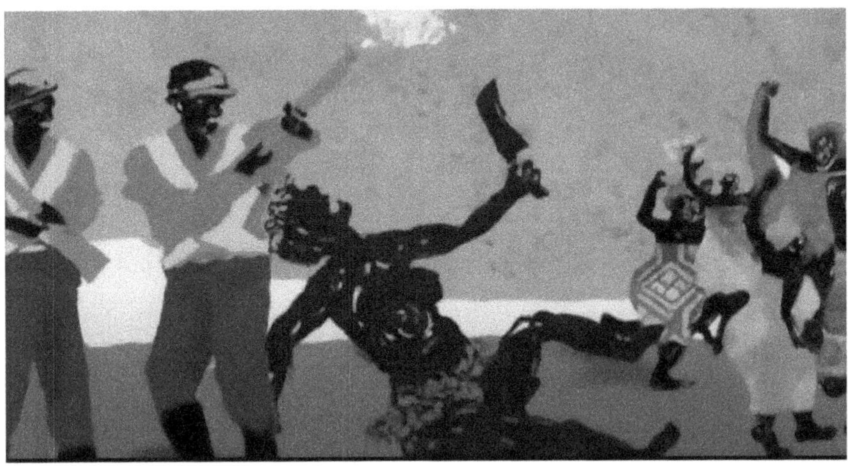

Figure 11. Artwork invoking the 1929 "Women's War" from *The Naked Option*, Deja Hsu. Dir. Candace Schermerhorn, 2011.

fetishized female body and weaponizing it: "We'll go naked for them. We'll do our naked. Because you people (Shell) want us to suffer and we're not going to take that. Fear will come" (*Naked Option* 00:01:32).

The petrofeminism of the women featured in *The Naked Option* uses the female body to expose Big Oil's damaging extractive practices but equally exposes those same bodies to gender-based violence: "At the risk of being raped, beaten or murdered—the women are prepared and armed—but not with anything you can see" ("The Women of the Niger Delta"). The documentary's subtitle, "a last resort," references a type of stock market trade, wherein, "with no protection from price volatility, such positions are considered highly vulnerable to loss and thus referred to as uncovered, or more colloquially, as naked" (Scott). The women share the vulnerability of the trader's position; both are engaged in risky business, trading without any underlying assets to fall back on.

The women are armed with nothing we can see. We cannot see weapons, because they are not armed with those,[5] nor do we see the nakedness with which they are armed. This is both because we are not the intended target of that threat and to protect the film's female subjects from any "epistemological violence" (Diabate 249) that might accrue were we to see them actually naked. Deja Hsu's stylized illustrations, used throughout the film, create a kind of

---

[5] Despite one of the many glaring errors in the film's official transcript, which states that the women of the 1929 war came armed with "pistols" instead of "pestles" (*Naked Option* 00:32:12).

scrim between us as viewers and the women's real bodies, representing nudity without showing it. The art connects the fight against Chevron to the historical Women's War of 1929, an action taken in protest of restrictions on female participation in local government that resulted in colonial authorities granting better representation. One tactic used in what is also called the Aba Women's Riots was nakedness, so the illustrations situate present-day protests in a specifically gendered lineage of struggle.

Frozen in Horror

In the sequences using the illustrations, *The Naked Option* cuts back and forth between the artwork and the live talking-head interviews. The art is animated in two ways: by the camera panning across still images, sometimes zooming in or out from certain figures, and through the use of cutout animation,[6] such as when the women's wraps fall away to show them naked. While made to move, the drawings themselves remain unchanged, which fixes the figures in place and fixes them in relation to the past.

In *Global Powers of Horror*, François Debrix asks, "How are we to make sense, if making sense is still possible, of the seeming shift from a violence that still cares about humanity, or, at least, about human bodily integrity to a violence that thrives in fragmentation, proliferation, and the pulverization of human shapes or forms . . . ?" (5). His answer (following on Adriana Cavarero's *Horrorism*, 2011) depends on distinguishing between terror, which "forces human bodies to run away in the face of violence and destruction," and horror, which "freezes" (5). Debrix's book is about geopolitical regimes of horror, not cinema, but the horror film does freeze things in a number of ways: It freezes characters in fright, viewers in their seats, and images into film frames.

The visuals for *The Naked Option* are full of images and clips of oil flares showing the horror of living in the contemporary Niger Delta, and its interviewees give voice to that feeling. Speaking about an oil spill, Annkio Briggs, executive director of the Agape Birthrights Organization says, "I recall how horrified we were . . . anytime we see a spill we're horrified because we can't believe that we can be seeing the same thing over and over and over again" (*Naked Option* 00:03:07). Briggs is horrified despite and because of the regularity of such disastrous events. It is not just the spills themselves but the witnessing of them that inflicts violence. Worse still is being caught in the fixed loop of repeatedly seeing them "over and over and over again." Similarly frozen in horror is the illustrated woman on the poster for the documentary. Caught in

---

[6] An early kind of animation often called "crude" in film histories.

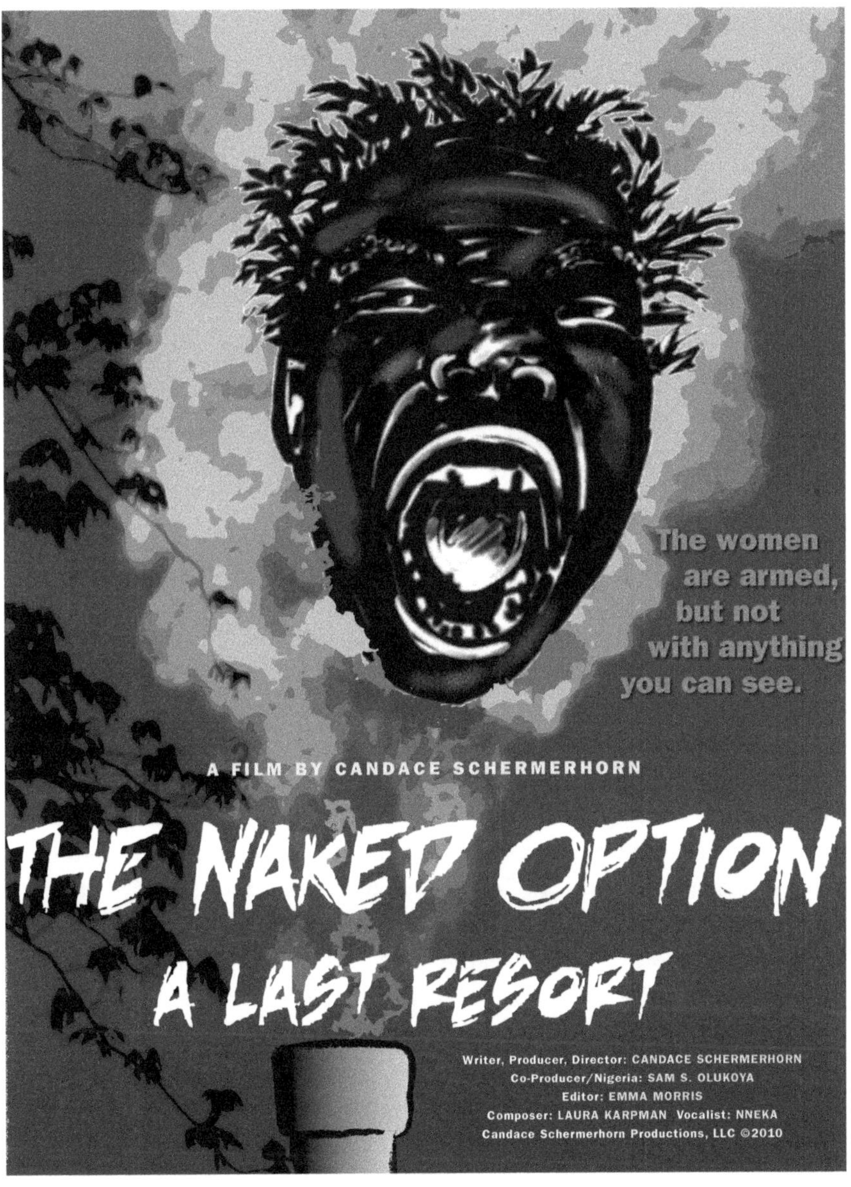

Figure 12. An enraged woman, her head superimposed over a gas flare, in the poster for *The Naked Option*, Deja Hsu. Dir. Candace Schermerhorn, 2011.

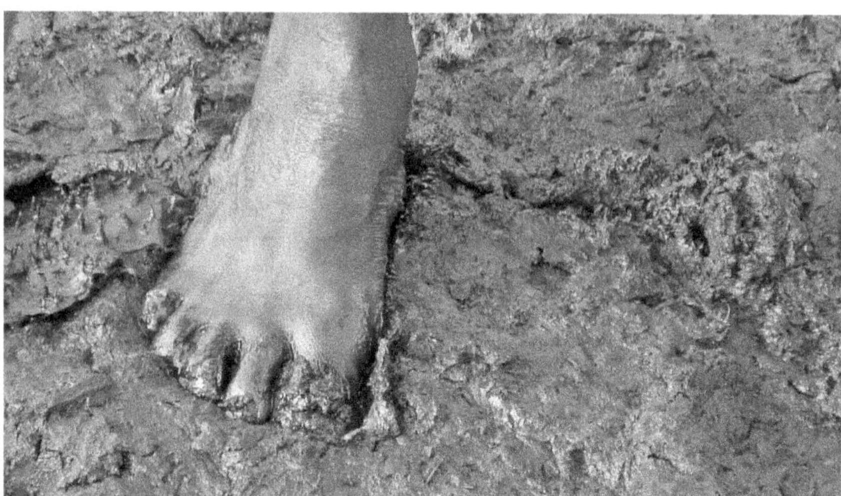

Figure 13. A woman's foot sinks into the contaminated mud of the Niger Delta. From *Daughters of the Niger Delta,* Dir. Ilse Van Lamoen, 2012.

a permanent version of *The Scream*, her disembodied head floats superimposed over the backdrop of a fiery gas flare, its reds and oranges an exact match for the "flaming clouds" in Munch's sky (qtd in Amos).

The films feature repeating images that freeze the moments in which human and nonhuman uncannily blur, transforming one into the other. A still taken from an early scene in *Daughters* unsettles because of the indistinguishability of the woman's toes from the mud, toes which look almost eroded by what we know is not just any mud but "a new type of matter," a toxic sludge of heavy metals, radionuclides, and other contaminants that can literally necrotize.[7] The surrounding muddy matrix threatens to suck her in, but meanwhile, she is stuck, frozen in place by the horrors all around her. Reading oil documentaries as horror films lets us see horror at work as "it renders human matter indistinct from nonhuman materiality (whether organic or inorganic)" (Debrix 6). Debrix ascribes intention to horror, describing "horror's own attempt at reforming or recreating matter—a new type of matter, perhaps—from the perspective of post-human, nonhuman, and inhuman fragmentation" (26). This is precisely the horror of oil, which takes on new forms itself, and in the process re-forms those who encounter it.

---

[7] See, for instance, a study that sampled the mud in order to understand the risks facing not only the local inhabitants but also the drilling crews in the Delta (Okoro et al).

The Living Dead

Transformation horror threatens the disintegration of bodily integrity and is a process "unpleasant to witness, and horrifying just to *imagine*" ("Transformation") because it renders unrecognizable what was once familiar. In a segment during *The Naked Option* discussing MEND, we hear that "by the time people become involved in that kind of activity, they might not be normal human beings anymore." Formerly "sons, or nephews or cousins or whatever," violence transforms them into the unfamiliar, even inhuman (*Naked Option*, 00:20:12). Shots of heavily armed and masked militants reinforce this impression—they are ominous and unnerving. The discussion ends, however, with a contrasting shot of a young boy standing in front of a corrugated metal wall graffitied with the legend "Trust No Body" (00:20:36). To trust nobody is a warning; to trust no body suggests that bodies cannot be trusted, not even the innocent-looking child who is certainly someone's son or nephew or cousin, since he may well grow up to be one of the "MEND boys" and therefore not a normal human being anymore. Next to the graffiti is a name—"Ibima"—which means "it will be good" in Ijaw, but we are left with the distinct impression that it will not be.

Despite the stakes, "the people believe very strongly that they have nothing to lose" (*Naked Option* 00:19:47); the conditions in which they live merit the tradeoffs of psychological and physical trauma. Put in terms of Mbembe's theory of necropolitics, they already inhabit a "death-world": "in our contemporary world, weapons are deployed in the interest of maximally destroying persons and creating *death-worlds*, that is new and unique forms of social existence in which vast populations are subject to living conditions that confer upon them the status of the *living dead*" (92). Mbembe's italics textually register the tonal gasp of dread with which we might speak these phrases that consciously reference the Holocaust but might equally be categorized as the language of petrohorror. Stella Fyneface, an activist interviewed in *The Naked Option*, names death unequivocally as a consequence of Big Oil's presence in the Niger Delta: "You know, before we became aware, they really cheated us to death, the companies. They cheated the communities to death" (00:23:02). After the credits roll, the film is dedicated to her memory. In her everlasting celluloid presence, she becomes the undead of the oil documentary, the living dead of petrohorror.

Wale Adebanwi and Ebenezer Obadare's collection *Encountering the Nigerian State* uses Julia Kristeva's *Powers of Horror* as a starting point for its claim about "state excess and the state constituted as an excess, the abjection these provoke and the consequences for citizens and groups within that abject state" (11). In their introduction and in later chapters in the book, such as Sarah Lincoln's on Ken Saro-Wiwa's novel *Sozaboy*, that excess gets figured—following on Claude Levi-Strauss—as "anthropophagic," "anthropoemic," or both, depending on the degree to which the state cannibalizes and consumes or vomits and excretes its citizens and adversaries. Lincoln's discussion of *Sozaboy* connects the excesses of the state to the excesses of the "excremental petro-economy on which that national project is constituted" (94); "Like any entropic system," she writes, "the Delta oil industry produces a remainder, an excess that cannot be absorbed back into the system. Along with variously toxic 'waste' products, which coat the surfaces of local soil and water and saturate the air, the process of oil extraction and global accumulation also produces 'waste peoples' whose superfluity to the global economy is systemic and irreducible" (81). The oil-state's excesses get talked about in terms of bodily functions because real bodies are laid waste to by the industry and because the physiology of emesis and excretion, the excess of the horror movie, aptly reflects the biopolitics of the sector. The excrescences of the oil state extrude in the vocabulary of horror.

Lincoln raises the spirits of werewolves (via Agemben 63), zombies (via the Comaroffs), vampires, and ghosts in her chapter in order to make sense of Mene's experience as a child soldier in *Sozaboy* and because Saro-Wiwa calls the nation that eventually murdered him "vampire-like" (*Genocide* 102), a designation taken up by Andrew Apter in *The Pan-African Nation: Oil and the Spectacle of Culture in Nigeria*. The "cannibalistic vampire state" (Apter 267) extracts ever-larger portions of human and extra-human resources. The accusation hearkens back to Karl Marx, quoted here by Jack Halberstam in a reading of *Dracula*:

> "British industry . . . vampire-like, could but live by sucking blood, and children's blood too." The modern world for Marx is peopled with the undead; it is indeed a Gothic world haunted by specters and ruled by the mystical nature of capital. He writes in *Grundrisse*: "Capital posits the permanence of value (to a certain degree) by incarnating itself in fleeting commodities and taking on their form, but at the same time changing them just as constantly. . . . But capital obtains this ability only by constantly sucking in living labour as its soul, vampire-like." (102)

New cultural and historical moments produce new monsters, according to Halberstam, and Marx's vampire-like capital suits nineteenth-century mass production and factory work. Sometimes late-stage "fossil capital" (Malm) sounds an awful lot like its predecessor, as in this description of life near the oil refineries of the Delta, which evokes images of industrial monsters: "If you are living in one of these villages, it was like staying at the back of a huge exhaust pipe that goes straight into your lungs and the lungs of children" (*Naked Option* 00:12:27). But, unlike modern industrialization, petroleum production does not need or want the labor of locals, instead exorcizing them into "deathworlds" as the undead of contemporary capitalism.

The Thing

In *Sweet Crude*, Chris Ekiyor, president of the Ijaw Youth Council, speaks: "Before the coming of oil, we had good fishes, good rich estuaries and good coastal land. We had no pipe-borne water, but we had fresh water that was floating, unpolluted, that our parents and our grandparents had, and we held it. They were just living and they were getting by. And then this thing called oil came" (00:15:35). In a conventional horror movie like *The Thing* (John Carpenter, 1982), "the thing" stands for something else: "The thing is nothing but a catalyst that brings out the monster inside all of us. Carpenter's film is about the internal conflicts that arise when paranoia penetrates the psyche; it is about what happens when suspicion disturbs the sanity of an entire group" (Khairy). In Ekiyor's telling, this "thing called oil" is animated and monstrous, but it's not anything other than itself. It's not a vampire or a cannibal; the horror of it is simply the thingness of oil and how it "came," unbidden, to the Delta.

Mbembe, Halberstam, and Apter's theories metaphorize the experience of being at the mercy of capitalist systems, but the documentaries make the metaphorical literal. The documentary film inherits these metaphorical monsters, but while theory and fiction have permission to lean figurative, the demands of the real on documentary mean it cannot accommodate such symbolic constructs. Instead, it exhibits the monstrosities of death and disease as its own logic of excess in a kind of documentary dissolution where the too muchness of the real spills over into something else, the ordinary horror of situations like the power going out.

The Power Goes Out

Number four on an online list of "The 8 Most Common Horror Movie Tropes and Cliches, Explained" is "The Power Goes out" (Vitalicio). When the power goes out during filming in *The Naked Option*, it is not because a villain has cut the lines or flipped the circuit breakers, the better to scare you in your own house as you "anticipate what awaits in the darkness" (Vitalicio), but because the power company can't be relied on for a steady supply.[8] The irony of power outages in a region of energy abundance is not lost on the women in the film or on the viewer.

The oil industry subsumes all other industries. "There is no industry in the Niger Delta. All there is [is] just the gas and oil because what they need to even sustain the industry which is the power is not there," explains Annkio Briggs in *The Naked Option* (00:09:22). It's important that the scene in which the power goes out happens inside someone's home. Because oil is the only industry and because Delta residents are excluded from participating in it, the public sphere of modern labor dissolves and gets swallowed up by the personal and the domestic. Composed of domestic entrapment, gendered violence, and being at the mercy of forces beyond oneself, ordinary horror is the petromodern equivalent of the female gothic and petrofeminism is what resists it.

Reflecting on the efficacy of the women's naked protests at the end of the film, Felicia Asekoni does not mince words: "Since then Chevron has not done anything for us. They promised certain things, then brought us another MOU [memorandum of understanding]. The one they signed before, they never did anything. Now they brought another one. It said 'global' and we don't want 'global.' We don't know what it means" (*Naked Option* 00:45:07). "Global" is yet one more unknown, and it, like oil, is everywhere.

Unseen Evil

In Zina Saro-Wiwa's four-minute, single-channel, looping video, *Niger Delta: A Documentary* (2015), part of her 2016 show, "Did You Know We Taught Them How to Dance?," we see the following shot from a stationary camera with which our eye aligns: In the foreground, on the beach, an empty red chair; in the middle ground, a lone fisherman punts his boat across the water; in the background, the opposite shore and sky. Ambient noise includes the wind on the microphone, the waves, and a rooster. In the very first and very last seconds of the film, what sounds like an accidental mic tap and the start of someone speaking lends authenticity to the impression of unstaged realism. Nothing else happens.

---

[8] The lights also went out during the filming of the Nollywood movie *Check Point*, as captured in *This Is Nollywood* (directed by Franco Sacchi): "Light in this country is not as constant as we would love it to be. I'll give you an example: we're just setting up now and the light went. But when we came here the light was on. So now we have to find a way of powering the background lights and all that" (*Nollywood* 27:05). I discuss the Nollywood film industry in the next chapter.

In her catalogue entry for the show, Stephanie LeMenager calls the film "a wry pastoral scene" in a bigger project about "what oil can't understand" ("Eden" 43). Is it wry? Or is it what LeMenager also says about Saro-Wiwa's work—that it presents "the Niger Delta as Eden in potentia" ("Eden" 40)? Wryness would undercut the simplicity of "a bucolic view of . . . peaceful waters" (Wilcox) and explain the film's ironic title: It is not what we expect of a documentary about the Niger Delta. LeMenager and Saro-Wiwa unironically[9] want it to be true that the Delta is not Paradise Lost, but that "'Ogoni is still Eden, only with pockets of Hell.' Not Eden if we close our ears and eyes, mind you. Eden if we dare to open them" ("Eden" 45). The video documents what we might not otherwise see, but by the same token it leaves out "what oil knows" (LeMenager, "Eden" 42). Calling it a documentary insists that we think about what it lacks that the other documentaries have—no voiceover, no action, no violence, no environmental nightmare. But those absences invite in the presence of oil. The horror is still there; we just can't see it. It sits in that empty red plastic chair, a thing made of petrochemicals, and waits.

---

[9] One more irony: *Niger Delta: A Documentary* has been acquired by the Chase Bank art collection. JP Morgan Chase made headlines in 2019 when it was sued by the Nigerian government for its role in helping high-ranking officials loot money from a lucrative license to drill oil in the Delta twenty years ago (Flitter).

CHAPTER 5

PETROCINEMA

Nollywood: "Where the Magic Lives"

In Franco Sacchi's 2007 documentary, *This Is Nollywood*, Bond Emeruva, a Nollywood director in the midst of filming *Check Point*, has a prescription for Nigeria's film industry: "Filmmaking," he asserts, "in an economy like ours . . . should be restricted to what I call edutainment" (9:11). He goes on: "So while we are entertaining, we should be able to educate. I believe in the power of audiovisuals. I mean, 90% of the population watches Nollywood. I believe it's the most valuable vehicle right now to pass the information across and to educate the people. So, if you're making a movie, no matter what the topic is, put a message in there" (*Nollywood* 9:29). In *Check Point*, his message is about corrupt policemen. For the petrofilms discussed here, the messages are murkier.

Nollywood is where Nigeria meets Hollywood and Bollywood, and as a film industry, it has been called "the most prolific movie machine" (Murdock). By far the majority of Nollywood productions are made quickly and cheaply for speedy dissemination via home video formats. Historically sound and image quality have been poor, production values low, and scripts heavy on dialogue and light on cinematic mise-en-scène (Iwowo 97).[1] Nevertheless, as Emeruva says, "This is Nollywood. Where the magic lives" (*Nollywood* 40:45).

Nollywood's specific sort of filmic magic[2] derives from the sheer volume of movies made[3] and the vastness of its viewership, who are transported via video

---

[1] For a summary of the film industry in Africa, see James Genova's "Film, Radio, and Society in Colonial and Postcolonial Africa."

[2] One more way in which Nollywood makes magic manifest is in its heavy indulgence in what Lindsey Green-Simms (in her essay with the same name) calls "occult melodramas" with their emphasis on magical money.

[3] Estimated at 2,500 movies per year, according to the Nashville Film Institute.

to aspirant scenarios in "a dream world where cities look and function like ide-
alized images of Los Angeles, Las Vegas, or Dubai" (Green-Simms, *Postcolonial
Automobility* 137). Enabling and underpinning this magical transportation
is oil, which film has always depended on to work its magic, from the magic
lantern, illuminated by oil lamps, to the invention of celluloid, a petroleum
byproduct, to the contemporaneous origins of the oil and film industries, as
described by the editors of *Petrocinema* in their introduction to that volume
(Dahlquist and Vonderau 1–2).

Indeed, this is a colonial, and in particular, Nigerian origin story.[4] We learn
from Mona Damluji that after Shell produced the first-ever petrofilm in 1924,
it went on to make film integral to company operations, featuring "subjects
made possible by petrol and petroleum-based products: aviation, racing, fire-
fighting, disease control, and robotics engineering, to name a few" (154). In
Nigeria, Shell refined its use of film as "evidence for its claim of improving
educational and social welfare in oil-producing countries," via the colonial film
unit (Damluji 156). Brian Larkin, in *Signal and Noise: Media, Infrastructure, and
Urban Culture in Nigeria*, devotes a chapter to the country's mobile film units:
"These were educational teams created by the government to show a mix of
documentaries, newsreels, and pedagogical dramas intended to instruct audiences about the achievements of the state and educate them in modes of
health, farming, and civic participation" (77). Larkin does so to center a different history of the "constitution of cinema" (76). His argument is that early
Nigerian cinematic form was shaped more by politics than by commodities. In
fact, about health screenings, he says they were "wholly outside of a commodity structure; instead, film was constituted as an object through which political
relations between colonizer and colonized were constituted" (Larkin, *Signal
and Noise* 86). That may be, but the use of cinema as an imperial tool paves the
way for industry films, like those made by Shell and other companies, to come
in and capitalize on a viewership grown accustomed to interpellation through
film. It also overlooks the constitutive relationship between the commodity of
oil and film. Recognizing the fundamental association between Shell and the
art of cinema (Damluji 154) makes it very difficult to place mobile film unit
screenings outside of commodity structure. As Damluji puts it, "the history of
cinema is a history of oil" (147).[5]

---

[4] In addition to Larkin, see Rudmer Canjels for a detailed history of Shell films in Nigeria.

[5] Not to mention that the very mobility of the mobile units (frequently used as other kinds of
conveyances, like taxis [Larkin, *Signal and Noise* 96]), depends on it.

In Northern Nigeria, the mobile film units were called *majigi*, after magic lanterns (Larkin, *Signal and Noise* 87). Cinematic magic played a big role in both designing and pitching the product, according to Larkin. Colonial filmmakers, speculating wildly about the information literacy of their African audiences, tried to rethink the "formal conventions of cinema," getting rid of pans, fades, montages, and more, as they would "only divert . . . attention from the scene message to the mystery of seeming magic" (*Signal and Noise* 109). Magic was used equally to mystify and demystify the technology. On the one hand, Larkin reminds us that, per Heike Behrend's research, "colonialists self-consciously presented electrical technologies like cinema to . . . audiences as a form of magic and wonder, a mode of power that could compete with indigenous supernatural force" (*Signal and Noise* 93). On the other, Larkin gives us this anecdote:

> One African commentator, writing in the magazine *Colonial Cinema*, explained that he started every show by saying to the audience, "I am sure you think this is magic," before giving a detailed lecture on how the camera worked, how film stock moved through the projector, even how film rested on the physiological illusion of the persistence of vision. (*Signal and Noise* 91–92)

Making the magic transparent meant that their viewership was being trained in the disciplinary techniques and messaging of Western modernity that would render them good subjects, productive workers, and eager consumers.

Oil itself gathers around it a magical lexicon often to do with visibility and visuality: whether it can be seen or is invisible, whether it is on display as spectacle or hidden in obscure depths. This tendency is true even of those who are most thoughtful about oil. For instance, while the Petrocultures Research Group, in *After Oil* write, "The reason . . . that oil modulates everything is not some natural or magical property of the energy source itself" (17), the "fairytale" they tell about oil ultimately reduces it to a magical thing:

> We envisioned the image of transition to "after oil" as partly an issue of visibility that we approached in terms of an archetype. We asked ourselves, if oil is the personal unconscious of modernity, then how do we make unconscious energy visible? Narratives and visual narrative form can make something visible and make the unconscious conscious so that we can grasp it and perceive it. Thus the fairytale of transformation can be seen as one archetypal narrative that captures the potential magic of oil and its

transformative power, acknowledging oil's seductive qualities. Oil is the magic that powers modernity. The power of oil is unconscious; we cannot grasp it and we don't perceive it. (47–48)

In contrast, the Nollywood films we are about to look at consciously try to make us perceive the power of oil.

Defining Petrocinema

While Marina Dahlquist and Patrick Vonderau's *Petrocinema* gives us a valuable overview of "the relation of cinema to the history of petroleum extraction" (4) it does not actually offer a definition of its titular term. It largely pays attention to "the dominant forms and formats of petroculture's cinema" (6) through the oil industry's sponsored films, designed to promote and assimilate their products into an emerging modernity. These films, like our Nollywood ones, also fall "outside of the domain of conventional cinema" (per the publisher's page). The industry films "enact forms of micropolitics that worked toward the goal of naturalizing the consumption of oil as energy and product, and thus to commodify oil" (3), while the films selected for discussion here intend to expose and excoriate the oil industry. In fact, they end up doing both.

Jeta Amata's 2012 *Black November (BN)* and Curtis Graham's 2015 *Blood & Oil (B&O)* are outliers to mainstream Nollywood in a number of significant ways. With far higher production values, goals of theatrical release, bigger budgets, and aimed at international audiences, they tick the boxes for being classified as new or neo-Nollywood productions.[6] Both showcase white oil executives as central plot points. And both are outliers in that they deal directly with oil at all.[7] Lindsey Green-Simms, discussing Nigerian video-films in her book on West African car culture, *Postcolonial Automobility*, observes that

---

[6] Jonathan Haynes gives us the term's provenance:

> Around 2010 the phrase "New Nollywood" began buzzing in Lagos and other places where people talk about Nollywood. The phrase describes an attempt by independent producer/ directors to "take Nollywood to the next level" by making better films with bigger budgets, films that can survive the aesthetic and technical challenge of being projected in cinemas rather than being released immediately as VCDs (video compact discs, the standard medium for movies in Nigeria) or DVDs for home viewing. ("'New Nollywood'" 53)

[7] There is a handful of Nollywood films dealing with oil, including *Liquid Black Gold* (which also seems to go by the name *Niger-Delta Avengers* [2008]), *King of Crude* (2011), *Crude War* (2011), *Militants* (2007), *Oil Village* (2001), and *Oil Money* (2003).

what also remains noticeably absent in Nollywood films is the commodity of oil, a commodity that literally makes the Nigerian economy (as well as the automobiles) run. With a few exceptions . . . films tend to ignore the devastation of the Niger Delta and to leave out any mention of the extraction of oil [or] the buying and selling of crude. . . . For the most part we only hear of oil when one hopes to attain a job at an oil company . . . or when oil contracts become fodder for 419 scams. (159)

Surely one of oil's unique qualities is this ability to be "noticeably absent." These two films tap into the power of the otherwise unseen, and, as petrocinema, make it visible and present.

The films were widely panned by critics, but it is precisely for that reason that they make for such useful texts. As with the short stories read in Chapter 1, they break out of storytelling mode to intervene in other ways. *BN* was criticized for being "filled with declamatory speeches, stereotypical characters and heavily telegraphed, melodramatic plot developments" (THR), for "being devoid of subtext," and for being "more a manifesto than a movie" (Rabin). The reviewer from *Nollywood Reinvented* said, "There is no room allowed for guesswork" ("Black November"). The same site slammed *B&O* (originally titled *Oloibiri*, on which more later), saying, "without the connection to the audience it ends up feeling like a really expensive sermon" ("Blood & Oil").

It is easy to dismiss the movies based on this kind of reception. Indeed, even Samantha Iwowo, the scholar who wrote the screenplay for *B&O*, complains about what she calls "illiteracy in neo-Nollywood" (100), where she believes there is a "disregard for cinematic vocabulary" (99). Although she dismisses the idea that Nollywood films have "fashioned their own grammar" (98), I argue that these two movies (one of which she wrote) are in their own category, *petrocinema*, formed (and re-formed) by their encounters with oil.

Petrocinema comprises films so steeped in oil that they reconfigure themselves as a result, sometimes proselytizing for it, as in the case of industry productions, sometimes sermonizing against it, as with documentaries, or sometimes a mixture of the two, as in Nollywood's petrofilms. Whereas in Chapter 4 I read oil documentaries as horror films, here I scrutinize popular feature films for the ways in which they are "factual, realistic . . . based on real events or circumstances, and intended primarily for instruction or record purposes" ("Documentary"), in keeping with how Nollywood filmmakers in general think of their art as edutainment and in step with Nollywood's place in a lineage of morality tales.[8]

The Funding Effect

"Really expensive sermons" they might be, but the politics of both films are ambiguous enough that it is not always clear what they are sermonizing about. *BN*'s ambiguity lies mainly in its internal politics—cinematic choices regarding whose point of view we align with, for instance—while *B&O* clumsily attempts to strike a compromise in its outward stance toward oil politics, for which it has been critiqued for leaving out the government's role, for taking oil money, and for representing locals and global corporations as equally culpable (Ekpo).

Funding streams for both films serve as indices of the oil industry's interpenetration with the film industry. Captain Hosa Wells Okunbo, an "oil baron" and key player at the commercial end of the petroleum industry for three decades, who, by his own account, did not want to go down in history as "just another rich merchant" but wanted to "give something back" to the region that made him phenomenally wealthy, spent $22 million on producing and promoting *BN* (Africa Report) and is listed as an executive producer. The film premiered at the Kennedy Center on May 8, 2012, and was screened during a United Nations General Assembly. *B&O*, in turn, received backing from the Nigerian National Petroleum Corporation (NNPC) (Ekpo) and premiered not just at any old venue, but at the Shell Hall in Lagos's Muson performing arts center on October 21, 2016.

When researchers distort results or modify conclusions due to undue influence from a study's financial sponsor, it is known as the *funding effect*,[9] visible here in the distorted forms of these feature films. Critics were not shy about pointing out the structural flaws in both films. *BN*, basically a remake of 2011's *Black Gold*, was reshot with 60 percent new content to make it "more current," resulting in a "crazy-quilt construction" in which "the seams of the film's bizarre production are all too apparent" (Rabin). In *B&O*, NNPC money

---

[8] Scholars differ on the extent to which Nollywood today is about moral messaging. Although James Genova pegs Ghanaian straight-to-video low-budget morality tales as a starting point for the industry, he goes on to say that "unlike the earlier decades of African cinema, which appealed to broad audiences and sought a general African consumer, the video films catered to communal identities and did not suggest a development project to the viewer" (16–17), whereas Green-Simms reads Nollywood productions as very much about "codes of morality" (*Postcolonial Automobility* 160).

[9] As originally identified by Sheldon Krimsky.

has birthed a hybrid progeny, "part action thriller, part historical eye opener, part government propaganda" (Ekpo). Whereas *BN* repeatedly shows Nigerian police acting at the bequest and on behalf of the oil company, *B&O* panders to its funders and, in addition to leaving out that militarized symbiotic relationship, spreads blame for the status quo evenly between village elders and industry players. The not-so-subtle implication is that failure on the part of the elders to protect the community in the past has led to disaster today. The main character, Timipre (Olu Jacobs), now an older man, is filled with guilt and remorse, and his juniors openly blame him: "I am the creation of your lousy generation," says the character Gunpowder (Richard Mofe-Damijo), now a militant (52:45). In their eagerness "not to offend but educate" (Ofime qtd Nwakunor), *B&O*'s filmmakers created something "not complete" (Izuzu) that "lacks cohesiveness" ("Blood and Oil").

Reflecting their shifting forms, both films were renamed. *Blood and Oil* had been titled *Oloibiri*, for the place where oil was "discovered" in Nigeria in 1956 by Shell D'Arcy. The film's replacement name, for the international market, is arguably more self-explanatory, though endlessly interchangeable (easily confused with the novel *Oil on Water*, the American movie *There Will Be Blood!* or a 2010 television series, also called *Blood and Oil*). Changing the name of something changes its meaning as well—we no longer understand its original intent,[10] in this case, a site-specific referent to the origins of the Nigerian oil industry. Now a free-floating, generic appellation, it might apply anywhere, at any time.

Black November's new title refers to the month in which author and activist Ken Saro-Wiwa was hanged by the state for his resistance to oil extraction in the Niger Delta, and the movie tells a similar story of a community's struggle against the corporation-nation alliance. Our protagonist is a young woman named Ebiere (Mbong Amata), newly returned to her village from being educated abroad on an oil company–sponsored scholarship, much like Saro-Wiwa's own female protagonist in the *Forest of Flowers* stories returns home as a symbol of progress. Like Saro-Wiwa, Ebiere is accused of masterminding the murder of tribal chiefs and martyrs herself for the cause. Unlike Saro-Wiwa, she gives love as her explanation, not the struggle, though she has become ever-more invested in it over the course of the movie. This softens the political message and also allows for the birth in prison of her son, clearly predestined to carry on his father's mission. Ebiere is executed on the same day as Saro-Wiwa was, November 10. Deservedly or not, the filmmaker, Amata, appears to liken himself to the famous activist. Having grown up in the Delta

---

[10] With thanks to Angus Parkhill for helping me think through this and other points.

and consulted with local militants about the film, he declared, "What I'm doing is what they did, but without weapons" (Africa Report).

Framed

Despite having quite different politics, the two movies have much in common, and are frequently lumped together as though they align "for one cause" (Ugochukwu 13). Local Niger Delta residents are up against Big Oil—under the all-purpose name Western Oil in *BN* and the barely-concealed Lesh (Shell backward, sort of) in *B&O*. *BN* begins with an opening slate of text on a black screen:

Nigeria. Population 167 million. World's 7th most populous nation. Over 521 recognized languages. Ex-British colony. English is the official language and spoken throughout. 90% of the population live on less than $2 a day. World's 5th largest exporter of oil. Continuous oil spills make it the most environmentally devasted land in the world. Average life expectancy: 47. (0:00:29)

In *B&O*, textual slates also bookend the action. These informational setups mean that we must understand the fictional narrative within the parameters of the factual frame narrative. Although the films go on to give us standard establishing shots, they cannot dispense with framing glosses. *BN*'s Amazon Prime video page, where you can stream the film, tells us it is "inspired by real events." *B&O* (also streamable) similarly derives its authenticity from its basis in reality: "*Oloibiri* was shot on location in the town of Oloibiri, like a documentary" (Saed). *Like* documentaries, but not, the films are full of oration, such as when, pretty much apropos of nothing, hostage-takers in *BN* declare, "My people are dying!" "Our lands are devastated!" "Our farmlands, livestock, wildlife" (*BN* 0:09:43).

Imre Szeman, reading oil documentaries "as providing examples of narrative and aesthetic choices through which the problem of oil is framed—or can be framed," quotes Fredric Jameson who says that cultural texts or artifacts are "symbolic acts" in which "real social contradictions, insurmountable in their own terms, find a purely formal resolution in the aesthetic realm" (424–25). I read our oil movies in opposition to that impulse, for the ways in which oil escapes the frame. The films, rather than finding formal resolutions, frequently break conventional form, reverting to documentary-like modes even within the storylines. In *B&O*, this manifests nostalgically as black and white flashbacks to mid-twentieth-century scenes of a funeral, activists organizing, and pastoral village life, with the palette implying found footage or old newsreel.

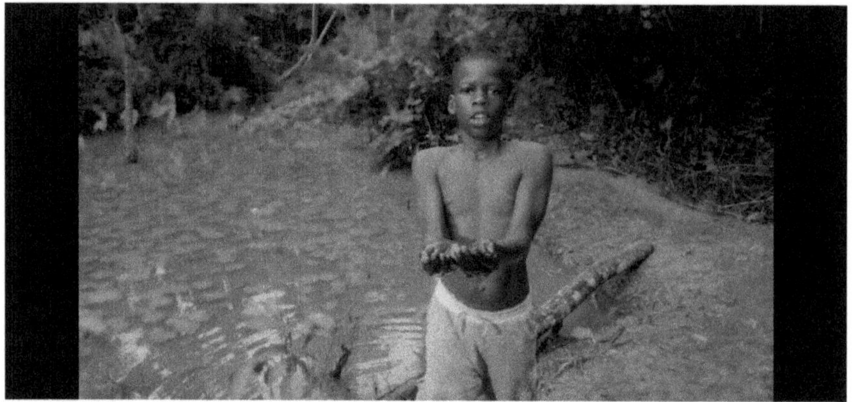

Figure 14. A young Timipre holds out hands full of oily water in a screenshot from *Blood & Oil*, Dir. Curtis Graham, 2015. Rightangle Productions.

Trigging one of these flashbacks, early in the movie, a female psychologist asks Timipre, "What hurts you the most in all this?" (0:07:48). We then see him as a lad, reaching into a pond, and, with cupped hands full of oil, turning directly to the camera in supplication (0:07:58). What hurts him, above and beyond the personal or psychological trauma he consults her about, are the social and environment conditions all around him. "Oil spills. Exploitation," he answers (0:08:00). The hurt felt is to the community, not to him, although individual bodies are also at risk of harm, which the film is eager to explain: "You know, the affected rivers have dangerous levels of cancer-causing agents" (0:40:18). Here documentary-like exposition does precisely what it means to do; it exposes. "In writing the script . . . I injected these truths," Samantha Iwowo, the screenwriter tells us (qtd Ajeluorou).

The desire for Nollywood oil films to bear witness to the Delta's degraded waters, desolate villages, and devastated terrain echoes the expectations we have of documentary film:

Throughout the history of film theory . . . there is an insistence on the capacity of film to record what is otherwise inaccessible to vision, opening up reality to that quotidian experience which cannot help but miss reality's full ontological presence and depth. One should not disavow the capacity of documentary to bear witness to reality in just this way, at the level of both form and content; the sublime of oil culture that these films visualize does not readily appear to everyday experience, which is one of the reasons why the consequences of the end of oil are neither feared nor acted upon. (Szeman 436)

We often think of the slow violence of ecosphere destruction in general and oil specifically as "inaccessible," invisible, or obscured, but the films both verbally explicate it and visually expose it. At one point, Timipre stops a boy from drinking river water: "He wants to drink crude oil" (0:24:40), he says, sarcastically. The visual parallels—like blobs of oil floating on the surface of the pond—are what Susan Schuppli would call "dirty pictures," which she explains are part of a new aesthetic of industrialization and its contaminating processes: "Anthropogenic matter is relentlessly aesthetic in throwing disturbing material re-arrangements back at us: dirty pictures of dramatically warped landscapes and polluted atmospheres that both intoxicate and repulse" (190).

In one long scene, the wide, brown river takes up almost half the screen, the oily water serving as a second screen within the shot, reflecting on its surface the forest on the opposite bank. Schuppli explains that oil has its own filmic properties: "While analogous to the workings of the cinematic apparatus, the oil spill is perhaps better understood as engaged in the production of a new form of cinema organised by the found footage of 'nature' itself" (193). We know—because we've been told—that the river is full of oil, so we are essentially watching an oil film within an oil film, "the actualisation of a ruinous image" resulting from "the conditions that brought about the disaster" (Schuppli 192–93). While Szeman reminds us of "the capacity of documentary to bear witness to reality," Schuppli asks us to consider that matter itself acts as a "material witness" to the "radical environmental transformations taking place all around us" (203), archiving their interactions with the world (206). *B&O* wants to capture the real, but Schuppli implies that if we train ourselves to see differently and on a different scale, we may not need petrocinema to bear witness for us. If we can apprehend "toxic ecologies as fully realized aesthetic agents" (194) (like the "image-making capacity of an oil spill" [193]), then perhaps it will no longer seem like "the sublime of oil culture . . . does not readily appear."

BN uses a different mechanism to achieve verisimilitude; the plot device of a series of television reporters and cameramen filming the action. Cameras filming within the film double the gaze on the characters, who become, by virtue of being interview subjects, interpellated as real activists, militants, and so on. They also place us in a viewing bind as multiple points of view compete for our gaze, complicating our positionality. We share the look of the camera as we watch the action, including watching other cameras film; we occupy the gaze of the audience inside the text when we see shots of television news clips; and we are recipients of the direct gaze of characters who look at us out of the frame. When the hostage takers go live on air, the camera cuts back and forth between extradiegetic direct address and United States government officials watching the broadcast, equating us with them as viewers: "Fifty percent of our oil comes to the United States. One out of every five Americans uses Nigerian oil. We export crude oil to you people, only to import refined oil. Why? Because Western Oil and our corrupt government wouldn't allow our refineries to work!" (*BN* 0:09:05) Thus we are held within another potential fiction—that of being international watchers, appealed to via privilege and power—regardless of our true selves.

Although the cameramen throughout *BN* are, indeed, men, the reporters are all (mostly white) women. In this way, the film skirts their female gaze (although they are the ones with access and whose explanatory voices we hear, sharing facts about the Nigerian oil industry) and subjects them to the male gaze as well. There is so much pseudo-reportage in the movie that the storyline sometimes seems only to serve as connective tissue, linking it together. About an hour into the movie, a minute-long sequence of global newscasts layered over village protest scenes (1:05:03) implies that an international audience's attention has been grabbed. Permission for the fictional plot to be hijacked by the facts of the matter is granted early in the movie, when correspondent Kristy Maine (Kim Basinger), caught up in the hostage situation, asks, "Do you want to tell your story? Do you want to tell the world?" (0:11:25). In this and many other moments, journalists act as accelerants to the already (literally) incendiary events in *BN*. Although they express ambivalence about their role (and one gets actively involved in the protests), it is nevertheless their insatiable thirst for the story that gives us access to it. Explains Kristy Maine: "It's a gold mine . . . the camera right on it, on everything. There's never enough" (0:19:05). They drum up interest, creating an unquenchable demand for content that mirrors the demand for oil, similarly manufactured.[11]

Both Nollywood/Hollywood collaborations frame whiteness at their cores. In addition to the conceit of white women reporters, designed to appeal to an

---

[11] Cara Daggett reminds us that "on the supply side, and especially prior to the 1970s' oil crisis, the state helped to secure an artificial oil scarcity that ensured profits for oil companies. On the demand side, the state also helped to cultivate oil desires, such that fossil fuel consumption became necessary to achieving the American dream" ("Petro-Masculinity: Fossil Fuels and Authoritarian Desire" 32).

international audience, they revolve around the hostage-taking of white male oil executives, both of whom escape intact. In *BN*, Mickey Rourke plays Tom Hudson, CEO of Western Oil and repulsive in every sense. While we are clearly not supposed to identify with him, he is nevertheless the reason we know about the Delta at all within the contours of this film. The *B&O* executive, on the other hand, objects that he set out to be "different" (0:18:55) in the Niger Delta and is distressed at seeing photographs of "people in profound pain" (0:18:32) mailed to him to literally expose conditions in the Delta. The camera lingers on the black and white pictures of emaciated children and ruined landscapes, including the ubiquitous image of hands covered in oil. But only briefly visible when he flips the first photo over is a message printed on the back: "Niger Delta asks: Shall we watch you come to destroy us some more?" (0:16:42). Part threat, part plea, the message returns and reframes the perpetrator's gaze (Baron).

The victims, however, are the ones who have been framed. Cutting away from the tormented executive, we get scenes of tacky opulence in his Nigerian counterpart's den of iniquity. Of a similar scene in a different movie, Green-Simms says, "the visual images of abundant wealth send a more ambivalent message: they place the conspicuous consumption of the urban elite on display and confirm the fact that there is indeed much money to be made in Lagos" (*Postcolonial Automobility* 134). The contrast between the two locations establishes Nigerian complicity with oil extraction and its consequences, a theme throughout the film that reaches a peak after that same executive (Powell, played by William R. Moses) has come to Nigeria to visit the "host community" (0:19:48), been kidnapped, and escaped. Derrida reminds us that hospitality, the duty of the host, is "a word which carries its own contradiction incorporated into it, a . . . word which allows itself to be parasitized by its opposite, 'hostility,' the undesirable guest [*hôte*]" (3). Framing Delta residents as the "host community" (which is standard industry practice, not just a throw-away line) imposes hosting status upon them, unbidden—not just for Powell's visit but for the whole industry—and hostilely demands hospitality, so that the "one inviting becomes almost the hostage of the one invited" (9).

Powell's status as actual hostage encapsulates all the contradictions of his presence there. *Hostage*, from the Latin *hospit-em*, also connects etymologically to *host* ("*host*age"), and Powell's presence opens up specific relational possibilities, extending him beyond himself: "The exercise of ethical responsibility begins where I am and must be the hostage of the other, delivered passively to the other before being delivered to myself. It is through the condition of being a hostage, says Levinas in 'Substitution,' 'that there can be pity, compassion, pardon and proximity in the world'" (Derrida 9). Now proximate, Powell unctuously exudes these qualities, but as ethical oil man, he is an impossibility.

The unsustainability of his position becomes apparent in the conversation he and Timipre have, walking along a riverbank, in which he manages simultaneously to assign fault to the inhabitants of the Niger Delta and to absolve his oil company:

> TIMIPRE: You siphon oil and send it to your country to build a better future. Oh yes, while our whole life is messed up. You bribe our chiefs so that they can keep quiet while you mess up with our oil and river. Huh! Forty years—for over forty years, Oloibiri, and indeed the entire Delta region, has been manipulated by people like you, and you know—you know—how far we have to drag to get proper drinking water.
>
> POWELL: I didn't come here to steal from you. I came here so my company can learn how to treat your community with respect. . . . Are your people completely innocent in this exploitation? It takes two people to screw . . . or get screwed. (0:59:53–1:01:30)

Having been put on equal footing, the men devise a novel outcome—the cancellation of the oil mining lease, which "no company has ever requested" before (1:01:47). As a coda to the story, they apologize to one another, sharing the blame (1:31:19). Easily missed, though, is the very last gesture of the movie: the Nigerian female psychologist mouthing "thank you" to Powell, signaling gratitude owed and tipping the balance of power once again back to him.

"The Only Language the West Understands"

Dirty pictures—like, for instance, "vast swaths of carbon-encrusted snow" (Schuppli 205)—might fill the need for new ways of representing the slow violence that is otherwise "low in instant spectacle but high in long-term effects" (Nixon 10). But Philip Aghoghovwia's entry, "Nigeria," for *Fueling Culture: 101 Words for Energy and Environment* immediately associates that country with spectacular violence: "One cannot write about energy culture in the Nigerian context without engaging the spectacle of violence it elicits, both in the public mind and in the sphere of the creative imagination, precisely because the form of sociality that oil energy generates in cultural production is imagined and inscribed in idioms of violence" (238). He offers two images for our consideration: the marker at Well No. 1 represents the "invisibility of Oloibiri" (239) and an American hostage at gunpoint represents "a quintessential image of the spectacle of violence" (241). They are odd choices; both feel selected with a particular "public mind" in mind—one looking from the outside in, one invested in a particular oil origin story and the threat to the familiar figure of the white hostage. In neither picture does oil show itself, although plenty of iconic oil images illustrate both subtle and spectacular violence. Where are the hands grimed with crude or the aftermath of a pipeline explosion?

To the extent that Aghoghovwia emphasizes violence while thinking of Western viewers, he would probably agree with the hostage-taker in *BN* who declares, "This is the only language the West understands" (07:27), while pointing a machine gun at Tom Hudson. The films are full of scenes of spectacular, graphic violence, including executions, torture, and rape. Their opening sequences set us up with expectations of violence to come—a voiceover bellowing "Murderers!" (0:01:47) during *B&O*'s funeral scene, and as *BN* begins, preparations for a hanging at Warri prison.

Cut to Los Angeles, California (*BN* 0:01:43), where, for almost five minutes, every shot is in or on a vehicle, capturing car chases, crashes, and general chaos, culminating in the hostage crisis in the 2nd Street Tunnel. Most of the American scenes in *BN*, in fact, take place in the tunnel, entombing us in an automotive landscape. The first dialogue we hear, between oil executive Tom Hudson and his adult daughter, takes place inside a car being chauffeured to the airport. "Daddy," she says, "You're always so stressed when we head back to Nigeria." "Yeah, well, nothing ever works, does it?" he replies (0:02:02). Cameras positioned inside vehicles let us see the interior automotive space being used as family room, meeting place, and interview location, much as Green-Simms describes when discussing the video-film *Blood Money* (*Postcolonial Automobility* 142). For the oil executive taking his work with him, the car becomes his mobile headquarters, signifying the reach and power of the oil industry, which extends from LA to Port Harcourt and back. Oil travels, and so does its corporate oversight. In *B&O*, executive privilege comes with a tourist experience as Powell is driven with armed escort from the airport to Oloibiri, passing on the way the signpost on the outskirts of town that reads, "Welcome to Oloibiri Town . . . Motto: the goose that lay [sic] the golden eggs" (0:36:27) and (over)seeing various scenes of impoverishment. It is a fine illustration of Green-Simms' principle that "the mobility of the wealthy and corrupt cause[s] the immobility of the ordinary people" (*Postcolonial Automobility* 142). Larkin calls the meeting (as a concept) "that arch form of corporate capitalism" (*Signal and Noise* 75), but it is worth noting that the safe space of the mobile office is breached in both scenarios, thus also breaching the capitalist contract, marked by "workplace practices and . . . forms of time, order, and technical ability" (Larkin, *Signal and Noise* 75).

"If automobility is about the liberal pursuit of personal freedom" (Green-Simms, *Postcolonial Automobility* 153), how does that tally with the use of the automobile as corporate space, a shared place of business? Firstly, because the oil company's raison d'etre is to privatize a public resource, thereby incorporating it into a regime of individual capital ownership. We see clearly in *BN* how the designation of "private property" (0:41:30) is enforced by the public sector in the form of state security forces who repeatedly attack and beat back locals protesting the oil companies' presence. Secondly, because the company men *are* their company, there is no legal or epistemological difference between them; even their families are subsumed by oil's plots—they are held for ransom or work for the firm. If the corporation is a person, so too is the person a corporation. Hudson says so outright: "I'm a corporate entity," he tells Ebiere (0:49:15). He can lay claim to this status because he has generations of corporate personhood backing him. He explains: "I was actually born in Nigeria. My father used to run a tin mine in Kaduna" (0:56:33). Compare his birthright assumption to the lack thereof for locals who scramble to gain access to the oil in their backyards.

Early in *BN*, villagers have rushed to collect fuel because the "pipelines are leaking again" (0:13:47). Just as we saw in the case of short stories in the first chapter, they are accused of "stealing" the "property of the federal government of Nigeria" (0:15:24). They only have access to it because of the leak, a leak that ignites one of many scenes of spectacular violence in both movies—a massive pipeline explosion (0:17:28). Oil leaking, spilling, being poured—oil in all of its excesses is the accelerant over and over. *B&O* begins with Gunpowder's gang dousing his rival's Mercedes in petrol and setting it on fire (0:05:04). (Excessive also in that cars are already highly flammable without the extra fuel.) Corrupt village chiefs who have collaborated with the oil company in *BN* are immolated in the same way.

It's no coincidence that automobiles feature so prominently in these moments of violence. Cars in general are central to Nollywood films, "both coveted and maligned, necessary and excessive at the same time" (Green-Simms, *Postcolonial Automobility* 127), and Mercedes in particular carry emblematic weight: "If one had to choose a single image to express the culture of the videos, it would undoubtedly be a Mercedes Benz, which appears ubiquitously as the symbol of the desired good life, the reward of both good and evil, the sign of social status and individual mobility" (Haynes, *Nigerian Video Films* 2).

B&O climaxes in a classic oil money shot: Gunpowder has been gunned down. A large canister of Tonimas' Enez Heavy Duty engine oil stands front and center during the shootout. Even in the scene that shows the catastrophic consummation of its effects, oil asserts itself in product placement. Now filmed from below, the canister tips over, and in a sequence of slo-mo shots, it glugs liquid out onto his prostrate body, covering him in spurts (1:29:29). As he

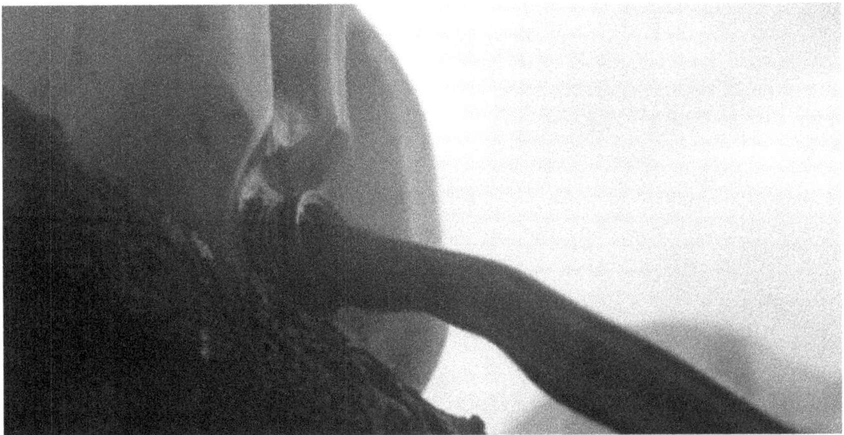

Figure 15. Oil spills from a canister onto Gunpowder in a screenshot from *Blood & Oil*, Dir. Curtis Graham, 2015. Rightangle Productions.

lies dying, we cut to his mother, who has a sympathetic response from afar, dropping a glass in shock. We hear her in voiceover: "I am Oloibiri. My son was pushed into militancy. Am I to be pitied as well? Has my son died in vain? I am Oloibiri. Never to let this tragedy happen again" (1:29:43). Fade to black. In this moment of peak oil, an affective kinship surfaces that transcends space and the individual, bringing us back to family and community, and giving proof to Daniel Worden's observation that oil continues to bind "to all family relations, even as oil's effect on the family is explicitly dramatized as pernicious" (443). The end of *BN* also shrinks the spectacular to the intimate. The threat of blowing up an oil tanker in the Los Angeles tunnel has been averted, and the explosion never comes. Instead, the final scenes show Ebiere being hanged, her baby and community outside the prison gates. Sometimes the language of petroviolence is spoken quietly.

The Narrative of the Country

BN's narrative closure comes predictably, foreshadowed by the noose shown at the beginning. If we know Saro-Wiwa's story, we also know how it ends. But what escapes is everything to do with oil: Not only is there no explosive finale, but Tom Hudson walks away, unscathed, to lead another day as CEO of Western Oil. Ebiere is dead, so who will lead the protest movement? What David Ingram writes about the very first eco-film applies here, too: "Although *Soylent Green* builds its narrative on environmental premises, therefore, its formulation of ecological crisis as already total, and of corporate and state power as monolithic, leaves little space for the formulation of a convincing politics of resistance" (155). The oil mining lease is cancelled in *B&O*, but the damage has been done. It's the exception to the rule, and we're reminded of that by the onscreen text that intervenes between Gunpowder's death and the final scene of reconciliation: "There have been over 2100 oil spills in the Niger Delta. There is still no potable water or fishing" (1:30:39). Abruptly brought back to reality, we realize that the fiction is held hostage by the facts and that even those who try to resist it are bound by petroculture.

Szeman, writing about oil documentaries, notes, "As is the case with documentaries on a wide range of social issues, these films about oil understand themselves as important forms of political pedagogy that not only shape audience understanding of the issues in question, but also hope to generate political and ecological responses that otherwise would not occur" (424). Nollywood's oil films feel the same pressure; they are expected to act like documentaries and give us the real. As a result, they break form as feature films, and spill out over their fictional narrative frames. They negate the Jamesonian assertion that aesthetic form serves as a container for social contradictions, tidying them up. The messy aesthetics of these films cannot resolve the contradictions of oil, not least of which is that it does not just live in the realm of capital, it is also a cultural and imaginative construction,[12] though still with brutal consequences.

This is true of our documentary-like feature films, as critics repeatedly hold them to a standard of producing real, measurable outcomes, a kind of reverse reality effect. Françoise Ugochukwu, for instance, thinks of "Nollywood as a sounding board and political advisory platform, screening a compelling summary of the issues, highlighting the stakeholders' conflicting interests, revealing the environmental and social damage and calling for solutions" (123). He ends his essay on Nollywood and the Niger Delta with a set of bullet points listing what the oil films have achieved, including having "faithfully and creditably captured the main issues facing the Niger Delta" (137).

The films did have real impact. *BN* generated a modicum of political response: American law makers sponsored a joint resolution aimed at pressurizing the Nigerian government and Western oil companies to clean up spills in the Niger Delta, titled, "Expressing the sense of Congress that as one of the world's important wetland and coastal marine ecosystems, the Niger Delta should be protected and its recovery and economic development [should be made] a priority" (United States Congress). (This was in 2012, so the degree to

---

[12] Per Peter Hitchcock reading oil both as a commodity and as a cultural logic ("American Imaginary" 81.)

which it had any impact remains to be seen.) As for *B&O*, Rogers Ofime, one of its producers, boasts that it garnered an apology to the community at its premiere (Nwakunor)[13] and helped gain approval for the Oloibiri Museum and Research Centre (OMRC), a project in development for thirty years, although it took until just recently, in early 2023, for the government to finally award a contract for its construction. The mission: "to convert the location where first oil was discovered into a monumental edifice that would preserve the heritage and developments in the oil and gas sector" (Udeme Akpan). Unsurprisingly, the OMRC is jointly funded by state and industry interests, including Shell, the Petroleum Technology Development Fund, and state government (Udeme Akpan). A statement released to the press touted the "socio-economic impact of the project" as including "employment generation, tourism, research & technology development and integration of oil and gas host communities into mainstream developmental narrative of the country" (Udeme Akpan). Perhaps it might generate some jobs, in a region with up to 64 percent unemployment (Chovwen), but how perverse would it be for residents to become promoters of the industry that ruined their lives and livelihoods? Once again, just like with museumized coal mines in the United Kingdom, or ex-prisoners leading tours on Robben Island, the tourist industry is poised to cannibalize other sectors. Plus, we know from experience that locals get passed over for work in the oil fields.[14]

The museum is, however, primarily a "symbolic project" (Udeme Akpan), a monument to oil that coalesces multiple and cruel ironies. But Oloibiri does not need to preserve the heritage of oil; that is already preserved in the loss of forest and agricultural land; the pollution of fresh water and fishing grounds; the declination of fish, crabs, mollusks, periwinkles, and birds; the destruction of large areas of mangrove forest over a wide area affecting terrestrial and marine resources; and the complete relocation of some communities, with the loss of ancestral homes (Kadafa). The heritage of oil has also been preserved in film. *Black November* and *Blood & Oil* show us oil as the natural state of affairs in the Delta—it has soaked into everything and is everywhere. It has become naturalized, just as those early industry films worked to "[naturalize] the consumption of oil as energy and product, and thus to commodify oil." And so, petrocinema, reflective of and in reaction to oil, ends up as one storyline in the "mainstream developmental narrative of the country."

---

[13] According to Samantha Iwowo, the screenwriter, the apology for "neglect" came from former Head of State, Retired General Yakubu Gowon (Ajeluorou).

[14] See, for example, Jennie Persson's study on economic opportunities for Niger Delta residents.

CHAPTER 6

PETRODRAMA

"Petroleum Resists the Five-Act Form"

Video killed the radio star and, according to Becky Becker, it allegedly killed theater in Nigeria, too. Writing in the journal *Theatre Symposium* in 2011, Becker declared the death of Nigerian theater at the hands of Nollywood, having visited the country and "left with the impression that the film and home video industry has had a disastrous impact on live theatre, despite its dependence on Nigerian theatrical tradition for its very existence" (69). Becker's essay establishes useful connections between theatrical and cinematic traditions in Nigeria, tracing how Nollywood grew out of traveling theater and mobile film units (as discussed in the previous chapter), and how the two forms look very like each other: "The videos often include just one camera angle (or very few), minimal cuts from shot to shot, and, in many cases, filmed-through scenes, much like a play that has been filmed for television or purposely created to preserve its theatrical qualities" (73). Becker concludes by suggesting that "when more film- and video makers begin to abandon stereotypical themes and stories and connect more keenly with the realities of the Nigerian people, theater will again find its voice" (77).

Certainly family dramas, the marriage plot, and folktales abound, but also a handful of Nigerian plays do engage with the realities of everyday life, and the particular matter of "urgent national interest" (Nnanna qtd Becker 77) of oil. In this chapter I turn to those petrodramas to ask how they handle the "volatile resource" that "burns through the[ir] scenes" (Clark, *All for Oil* 2), paying special attention to representations of women, gender relations, and female playwrights (although section two does look at a sampling of plays by more famous male playwrights to give the long view and to afford some comparison and contrast in approaches to the problem). As with all the objects of study in

Petroforms, I have only selected plays that explicitly involve oil, not those that might simply assume petromodernity as their backdrop.

More than one of the plays we are about to discuss have been called "Brechtian," picking up on their social purpose, their (often satirical) political messages, their episodic structures, and their use of parable. The label fits even more neatly when we remember that Bertolt Brecht himself uses petroleum as the example by which he thinks through form. First, he establishes that new relations are generated by new industrial processes and mechanical means of production, singling out petroleum as exemplary:

> The first thing to do, then, is to identify the new subject matter; the second to shape the new relations. The reason: art follows reality. An example: the extraction and refinement of petroleum spirit represents a new complex of subject, and when one studies these carefully one becomes struck by quite new forms of human relationship. A particular mode of behavior can be observed in both the individual and in the mass, and it is clearly peculiar to the petroleum complex. But it wasn't this new mode of behavior that created this particular way of refining petrol. The petroleum complex came first, and the new relationships are secondary. . . . [P]etroleum creates new relationships. (*Brecht on Theatre* 29–30)

Next, he enlarges on that premise with the idea that these new subjects and relationships *must* produce new forms:

> Simply to comprehend the new areas of subject-matter imposes a new dramatic and theatrical form. Can we speak of money in the form of iambics? "The Mark, first quoted yesterday at 50 dollars, now beyond 100, soon may rise, etc."—how about that? Petroleum resists the five-act form; today's catastrophes do not progress in a straight line but in cyclical crises; the "heroes" change with the different phases, are interchangeable, etc.; the graph of people's actions is complicated by abortive actions; fate is no longer a single coherent power; rather there are fields of force which can be seen radiating in opposite directions; the power of groups themselves comprise movements not only against one another but within themselves, etc., etc. (*Brecht on Theatre* 30)

Again, petroleum doesn't just stand in for industrial processes (or, indeed, for catastrophes in general), although it encompasses their full spectrum, from extraction to consumption, but *in particular* is the example that "resists the five-act form" of conventional playwriting. Posting on the social media platform Twitter (now X) in 2022, Shane Boyle wrote: "Reminder: when Brecht says that 'Petroleum balks at the five-act form' he is not saying that oil can't be represented but that bourgeois drama (plot-driven dialogue) isn't built to grapple with the infrastructural complexities of extractive industries." This underscores what we have seen throughout this book: not the failure to represent oil, but the challenge of how to do so. Boyle's slightly different translation is also telling: Where "resist" suggests possible override, "balk" indicates refusal.

Brecht, writing in 1957, is exactly contemporaneous with the start of production in Oloibiri's oil field. International headlines at the time read "Nigerian Oil Strike Made, Company Says" and "Britain Gets Nigerian Oil." Petroleum was on the public mind.

Bodies Corporate

Nigerian plays by men that "enact the aesthetics of the collective struggles against petro-capitalism" (Ajumeze, *Biopolitics* 63) include John Pepper Clark's *The Wives' Revolt*, Esiaba Irobi's *Hangmen also Die*, Eni Jologho Umuko's *The Scent of Crude Oil*, Isaac Attah Ogezi's *Under a Darkling Sky*, Uzo Nwamara's *Dance of the Delta*, Ahmed Yerima's trilogy, Oyeh Otu's *Shanty Town*, Chika C. Onu's *Dombraye*, and Hope Eghagha's *The Oily Marriage*.

Of these, two stand out for placing women at the nexus of oil money and gender relations: Clark's *The Wives' Revolt* (1991) and Eghagha's *The Oily Marriage* (2018), both by playwrights from Delta State.[1] As if to confirm Brecht's thesis that petroleum resists the five-act form, neither is in five parts. In *The Wives' Revolt*, the women of the village of Erhuwaren near the Niger Delta go on strike over the unfair distribution of a payout from an oil company drilling on their land. Self-evidently a revisitation of Aristophanes's *Lysistrata*,[2] the play explores the consequences of withholding—both financially and sexually. It opens with a long soliloquy spoken by Okoro, town crier and husband of

---

[1] Yerima's *Little Drops* from his trilogy also centers women but its attention is on the extreme, dehumanizing, largely sexual violence they experience as they are caught up in the Niger Delta conflict zone.

[2] As are Julie Okoh's play *Edewede* (2000), which deals with female circumcision, and Stella Oyedepo's *The Rebellion of the Bumpy-Chested* (2002), where women turn the tables to oppress men.

Koko, one of the leaders of the women's revolt. The strike is already underway, and the town is feeling the impact. Okoro speaks of the "women-folk":

> In pursuit of their claims, which we declare are not only preposterous but in complete violation of our ancient custom and law, they have refused for many a day now to perform their civic duties and responsibilities, as a consequence of which our streets, our public places, right to the market square, that is their own preserve, are today filled with the rank excrement of earthworms and goats, roaming unchecked in our city.

He goes on to accuse them, "by power of witchcraft," of assuming "the insidious forms and shapes of goats to terrorize" the town. Clark steps straight into the scatological; the women are said to be exercising their excremental faculties, creating a "condition of filth" and a "stench that hits the nose like a cockroach flying in your face." Whether by neglect in absentia or resolve in the form of goats, they have subjected the men who remain behind to unpleasantness. To cope, Okoro decrees that no one, "either as individuals or bodies corporate," may own a goat (Clark 1).

Koko is her own character, but the play for the most part treats the revolting women en masse, as a body corporate. Despite Okoro mocking them for not being able to unite ("You cannot even speak with one voice on any one matter at any time" [6]), scene three, "Walk-Out," proves otherwise when they collectively act. By walking out, the women of Erhuwaren take their place in a proud line of Nigerian women before them who have organized in protest, with the play explicitly referencing the Aba Women's Riots of 1929 (over the restricted role of women in government) and the Abeokuta Women's Revolt in the 1940s (against the imposition of unfair taxation) (10). Not mentioned but surely influential are the women's revolts in the 1980s against oil industry installations and personnel, a model taken up again in the 2000s during the "naked protests" in which women threatened to disrobe in order to shame oil workers (as previously discussed in chapter 4 on documentaries). In the play, the wives leave town, leaving the men to the dirty work of childcare and housework, and when they return, they have collectively caught a sexually transmitted disease, intentionally spread by the town's enemies via a notorious woman who has contaminated them without contact (40).

The disgust, blame, and shame that follow this discovery echo thematically through Eghagha's *The Oily Marriage* (*OM*) as well, a play in which oil mediates all relationships, rendering them "very oily" (46). It follows the simple plot line of two people who wish to get married, Odion and Maiden, but who are revealed to be half-siblings. Before they are made aware of their heredity, though, they have already committed an "abominable act" (46), a judgement reiterated in both plot and language when characters discuss a man who slept with "his pretty daughters" (78):

> ADJARHO: See how the man ended!
> AWOKO: In spite of his wealth, he died drinking his piss and eating
>   his shit!
> ADJARHO: Abomination. That is what happens to a man who behaves
>   like the he-goat and mounts his mother! (78–79)

The "excremental trope" (Esty 24) is everywhere in these plays, from the "filth" of the goats (Clark 42) to Okoro getting "wee-ed" and "pooed" on when he feebly tries to babysit his child (Clark 25). As a "discursive resource . . . representing postcolonial disillusionment" (Esty 26, 30), it does everything that Joshua Esty attributes to it: "shit acts as a material sign of underdevelopment; as a symbol of excessive consumption; as an image of wasted political energies; and as the mark of the comprador's residual, alien status" (34). The man ends reduced to his own waste products not "in spite of" but *because of* his wealth: He is one of the many male compradors in the plays who benefit, however scantily or tangentially, from the oil industry.

Sweet, crude oil is simultaneously the stuff of lust and disgust. As petroleum works its way through the economical digestive tract, it gathers dirty associations. There are the real conditions of extraction, refinement, and processing that incontinently spill into our environments and the metaphorical associations that accumulate as oil turns into oil money. In "Oil as Money: The Devil's Excrement and the Spectacle of Black Gold," Michael Watts explains how Perez Alfonso, founder of OPEC and former Venezuelan oil minister, came famously to call oil money "the devil's excrement": "In Perez' powerful and compelling vision, the natural bounty of oil had, in the magical and mysterious process of being transformed into money, become a putrid and toxic waste" (205). Both Watts and Stephanie LeMenager have written about "Oil's dirtiness and fecal qualities" (LeMenager, "Aesthetics" 92) as a "black and sticky fluid" (Watts, "Petro-Violence" 191). In the hands of a writer like Ayi Kwei Armah, that shit stinks: "He re-odorizes money, converting it into shit and forcing readers to see wealth as polished waste. . . . In a system entirely out of economic balance, shit flows through the novel like an alternative currency, a cruel displacement of productive capital" (Esty 33). On the other hand, Watts reminds us that the smell can be masked: "Oil money—as social power, as state corruption and degeneracy, as blind ambition and illusion—has eroded sociability, turning everything it touches into shit: oil money as deodorized feces that has been made to shine" ("Oil as Money" 219).

From the beginning, the theory of petrofiction has been redolent with oil's smell. In Amitav Ghosh's article that established the field, "oil smells bad" primarily in a figurative way, as an ethical "Problem" ("Petrofiction" 30). Henry Ajumeze, who also uses Esty to think through petrodrama and scatology, notes "the ways that postcolonial discourses are extended to implicate different forms of power formations—including the power of odor" (99). Ajumeze's text is Ben Binebai's play *My Life in the Burning Creeks* (2014), which supplies him with a narrowly masculine scatological reference point in the form of the male anus (because the narrator repeatedly refers to his). Here, however, we see the excremental feminine. It is women who create the conditions of filth and stench, and the smell of oil is distinctly sexual. Eghagha's author's note for *The Oily Marriage* starts, "There is something ferociously lusty about the smell of crude oil that attracts both men and women" (5). When the play then opens with Chief Adjarho, Maiden's father, "smell[ing] something in the air" (13), he's sensing a whiff of unsavory developments to come, but his wife assumes it's "the smell of oil money" (13), which he loves. There's "nothing wrong with loving the smell of oil-money which is under the soil of your house!" (14) she confirms.

Loving or hating the smell depends on whether it is yours or not, and the distribution of oil money, that "dirty lucre" (Watts, "Oil as Money" 218), follows distinct patterns. We know that even if it "is under the soil of your own house," you are likely to be divested of it. Additionally, these petrodramas reveal the masculinist, misogynistic division of profits. In *The Wives' Revolt*, Okoro's opening speech explains that the women's complaint is that men benefit twice over from the oil company's settlement, since the elders, allotted one third of it, are all men. Eghagha's author's note about the lusty smell of crude conflates sexual attraction and oil extraction, setting up his oily marriage subplot in which Maiden's father wants to marry her off to the son of a Flames Oil Company contractor. Eghagha goes on to equate women and oil as male infatuations (6), and gives his character a generic name that marks her, just like crude oil, as a virginal, desirable possession ("acquisition is impolite," Eghagha writes in parentheses [6], in an attempt to be feminist that backfires). Maiden recognizes her predicament. At the end of the play, after rejecting a ring and an employment contract, she monologues: "They all want or wanted or loved me for different reasons, some good some not so good. My father loved me because I was his gateway to national wealth, fame and politics. It's sad that my father whom I loved so dearly also commoditized me. . . . I am the oil of love, the one exploited" (102–03). The characters in the play understand themselves to be in a petromodern version of the traditional marriage plot—"Throughout history, there has always been a marriage between commerce, religion and family" (56), lectures one—with Maiden as its distillation. When she declares, "I am the oil of love," she gives voice to the commodity with which she has been identified.

Although the women in the plays suffer from chronic structural and cultural inequities, they nevertheless persist in trying to angle for better positions and payoffs. Chief Adjarho, Okoro, and the other male compradors are go-betweens for the oil companies, managing youth "protests against oppression, unemployment, non-indigenes getting all the juicy jobs" (Eghagha 37), divvying up remunerations, and passing decrees. The female comprador's position is more precarious. In *The Oily Marriage*, men and women join to protest the importation of "young men from across the Niger to come into our towns and villages to guard pipelines" (31), chanting, "*All we are saying give us our jobs/ All we are saying don't take our oil*" (37). "Female Youth One" shouts, "Monkey dey work baboon dey chop! Empowerment for the girls!" (37), meaning some people work while others get all the benefits. However, the abstraction of female empowerment does not hold up to the particularities of reality, in which Maiden's employment opportunity, as Community Relations Officer for Flames (73), is really only part of her father and other male authorities' "purchase plan" for her (51). Maiden's refusal to enter the oil industry via the marriage plot is also a refusal to become a body corporate, a wife arranged in the service of the company. Her final speech, which ends the play, is defiant but vague as a declaration of independence. The only thing it resolves is that the offer of employment should go to her brother, keeping oil in the family.

In Clark's play, it turns out the wives revolted over a paltry sum:

> As for the original matter of the oil company money that started all this fire, let it be known here and now that it was not such a big sum of money. Certainly, not so big that it was going to change the condition of our lives permanently for better. It has left the poor, poor, and the rich perhaps a little richer as our oil continues to flow to enrich other people across the country. But that is another story. (Clark 42)

The consensus to use the money to start a school building fund means the community in general is the beneficiary, not the women specifically. After announcing this outcome, Okoro appears to get the last word: "The story is therefore finished, it is finished!" (42), a line that parallels the ending of his opening speech. But the stage directions dictate that he start his lines all over again in the wings (which is also the instruction at the end of scene one). Despite the declaration of being finished, he cannot finish; he is caught in an unstageable infinity loop, a repeating form that mirrors the cycle of oil money he has just described.

One critique of Clark's characters is that he "seems to present the marginalized women in his plays as a metaphor for the peasants in the Niger Delta who suffer the brunt of oil exploration, and the male gender as the elites who appropriate the monies meant for the development of the region" (Okpahdah 54). This general tendency toward allegory results from Clark's lumping together of the "wives," and, in Eghagha's play, from Maiden standing in for young women in general. According to Brecht, it is also a product of trying to stage plays that show "modern processes," specifically oil prospecting and extraction. Here, he reflects on the dramaturgical difficulties presented by petrodramas: "E.g. there was a play called *Petroleum*, originally written by Leo Lania but adapted by us, in which we wanted to show exactly how oil is drilled and treated. The people here were quite secondary; they were just cyphers serving a cause" (*Brecht on Theatre* 66). In the generic repetition of characters, the repeating structure of Clark's play, and the perpetuation of oil relations in Eghagha's, we see that petroleum keeps repeating itself, resisting the five-act form.

The Strange Case of the Duplicate Plays

If you were to look up the playwright Tess Onsonye Onwueme, you would see, among her many credits, two plays about oil, one called *Then She Said It* (2002) and the other *What Mama Said* (2003). You'd be forgiven for thinking they were two distinct plays, what with two titles, two publication dates, and two publishers. You might then be surprised to discover that they are basically the same play, with only slight modifications. Certainly reviewer N. Graham Nesbith was:

> *What Mama Said,* at 200 pages, is indeed a long evening at the theatre for American audiences. More baffling is the fact that the dramatic theme and account are ostensibly the same as in the earlier *Then She Said It* (2002). The desire to rewrite and revise is understandable; playwrights do it all the time. New characters, plot twists, better dialogue, and altered dramatic premises are not unusual when playwrights update a script. Better proofreading and trimmed dialogue are the chief contributions of *What Mama Said.* There are no substantial changes from the original text to warrant a

new title. Two titles suggest two different plays, yet that is not the case here. (257)

In the strange case of these duplicate plays, it would be easy to suspect Onwueme of double-dipping, but it is far more interesting to consider the repetition as a "peculiar" (to use Brecht's word) symptom of our encounter with the petroleum complex.

Briefly, the plays' plot pits an increasingly angry "mob" (3)[3] of "wounded community" (2) members, made up of women and youths, against a handful of men with interests in the oil industry. Much of the action occurs off-stage and is reported to us in dialogue. On stage, a split set shows the GRA/Oil Club and a market square.[4] We see the men boozing and schmoozing as the crowd draws nearer and their demands, in particular for "resource control" (15), get louder. Eventually, a number of oil men, including the Offshore Oil Director (not shown) are kidnapped. The government arrests some of the suspects, others escape, and in the epilogue, the Oil Club transforms into the Supreme Court, where a trial takes place.

There is no difference in plot between the two versions of the play, and most of the language is repeated verbatim, but there are a few ways in which the plays diverge in addition to the edits I alluded to. Both take place in lightly veiled versions of Nigeria. We know this for a number of reasons, such as the use of Nigerian Pidgin English and a moment when a character starts to say Nigeria and corrects himself with the country's fictious name (43). In *Then She Said It*, this is the State of Hungeria; in *What Mama Said*, it's the State of Sufferland.[5] Next, the characters have been renamed. More on this in a moment, but the most noticeable change is that Ethiope, the Traditional Chief (a close ally of the government official and the foreign oil director) is now called the much more on-the-nose Pipeline. Also, what were scenes in the original are now called movements, and since one was divided, there are now twelve rather than eleven of them.

Last but not least, by adding the subtitle "an epic drama" to *What Mama Said*, Onwueme lays claim to the Brechtian tradition of epic theater. William

---

[3] Unless otherwise indicated, page numbers in this section of the chapter refer to the first version of the play.

[4] Where GRA = Government Reserved Area, originally for colonists, latterly just for the wealthy.

[5] By renaming Nigeria, Onwueme's play takes its place in "a whole raft of plays set in partially recognisable African states" (Edgar).

Over situates her in relation to both Brecht and Nigerian national drama, concluding that,

> In her dramas, the "alienation effect" (*Verfremdungseffekt*) is realized . . . through artistic innovation, social parody, allegorical names and characters, and a clear movement back from theatrical sentimentality towards social and political engagement expressed through an evolving solidarity among the characters (and audiences?). In this way she moves beyond the binary of distance and empathy to create a unique feminist theatre where feeling both for and with the characters occurs within contexts of historical change. (187)

Because Over reads only three of Onwueme's plays in this light, leaving out the oil plays, he overstates the realism and specificity in her stagecraft (177) and the degree to which she establishes empathic relations. Doubt about that, expressed in his parenthetical question mark, pertains to the petrodramas anyway, especially with regard to their final stage direction: "*(With spotlights on them in this call-response chant, the rousing drumbeats empower the women and youths into a huge orchestra as they shift out of the stage and into the audience, to build into a communal dance party until floodlights.)*" (221). Nesbith, in his review, calls this ending "expected" (257), suggesting less actual spontaneous political solidarity and more ritualistic display. As ritual, the ending gestures toward traditional forms "not to examine and privilege tradition but as means to ends, as theatrical devices in the service of politically based issues" (Over 186–87). It is also alienating in Brechtian fashion; a ceremonial ritualization that establishes distance.

The alienation effect manifests in various ways in *What She Said/What Mama Said*. Onwueme makes extensive use of the chorus, equally a Brechtian and an African move, with its call and responses sequences giving voice to the communal female, as in Clark's and Eghagha's plays: "While these questions and their responses delineate the women's personal experiences, the effect is a collective awakening to their shared reality, which helps to galvanize them into an oppositional body" (Eke 9). Alienating in the same way—because we don't form individual character attachments—is the fact that actors play more than one part, with costume changes taking place in the open:

> *The Young Woman quickly exits behind and returns with a box-full of clothing, house hold items and whatever else is required to dramatize their experience. The people are so animated and wired up that in no time, the items are all organized*

or distributed and they dress up for their new roles. Thus transformed, some begin to mime and adjust to their new characters/personalities, while others set up the stage. (3)

The choice of multiple roles is potentially confusing for the audience, but also seems deliberately designed to blur the distinctions between the characters' politics. For instance, a single actor performing the Traditional Chief, the Trial Justice, and Businessman draws attention to how they protect their shared interests. Maureen Eke agrees: "In collapsing the identifies of several figures . . . Onwueme succeeds in signaling a continuity in the hierarchies and relations of power" (10–11).

Eke writes this in her introduction to *What Mama Said*, another addition to the original. For Nesbith, "the introduction captures and provides what the play does not" (257). Nesbith assumes an American audience that is not already intimately familiar with the issues at stake (257), but regardless, the new introduction indubitably provides an explanatory exoskeleton, outside and in excess of the play itself. Nesbith's other structural critiques include the play's "myriad loose ends" and missing key scenes (256–57). Sometimes, events that get referenced in stage directions don't get shown to the audience, so they are missing in translation from page to stage.

But the most extreme departure from form is the duplication of the two plays. To take the duplication seriously is to see the play happen again, and with political implications. In Brecht's essay, "The Street Scene: A Basic Model for an Epic Theatre" (1938), he coaches us that "theatre for a scientific age" (1) must have all the elements of a scene on a streetcorner where an eyewitness replays a traffic accident: "The street demonstrator's performance is essentially repetitive. The event has taken place; what you are seeing now is a repeat. If the scene in the theatre follows the street scene in this respect then the theatre will stop pretending not to be theatre, just as the streetcorner demonstration admits it is a demonstration (and does not pretend to be the actual event)" (2). In our analogy, the second version of Onwueme's play is like the eyewitness's reenactment: in its repetition, it calls attention to its own performativity. Moreover, "to qualify as epic," according to Brecht, "the demonstration should have a socially practical significance. Whether our street demonstrator is out to show that one attitude on the part of driver or pedestrian makes an accident inevitable where another would not, or whether he is demonstrating with a view to fixing the responsibility, his demonstration has a practical purpose, intervenes socially" (2). In other words, the point of the repetition *is* social intervention. Onwueme's dual plays double down on calling attention to the scene of the accident, in this case, the discovery of oil and everything that followed, as outlined by the Defense Lawyer during the trial:

> For nearly half a century, Your Honor, the living mothers, sons and daughters of the land have been trampled. Oppressed. Exploited. And dehumanized. You're all witnesses. Since the Sweet Crude was discovered in their land, each one of them has had their own personal tragic experiences. They've been without jobs, farming and fishing resources . . . painful, hard experiences that progressively changed their lives for the worse. My fellow citizens, it is not for me to tell, for as they say, she who wears the shoe knows where it pinches. My learned friend, reinforced by his prosecution witness has presented these poor women and youths as the nagging pestilence and diseases afflicting this land. Now before you all, they stand. Helpless. Poverty stricken. Shackled. Mothers, sons and daughters of this land who have become prisoners of the state, foreign and private interests. Here they are. (211)

Specifically, Onwueme wants to call a global audience to the scene. Significant portions of the script grapple with how the rest of the world sees the situation in the Delta—the women want to draw "International attention to our cause!" (22) whereas the men fret: "International disgrace? And we're only just trying to clean up our messy image before the whole world?" (185) The repeat reflects the never-ending quality of the crisis, as articulated by the leader of the market women: "Our stories are so long that if we all were to speak and tell the world all that we have been through, our narratives would never end. So now, we can only share with you a bite. Just a small bite as evidence of what eats deep within us" (213). With repetition comes difference, and into the repetitive space between the two versions of Onwueme's play enters the possibility of intervention.

One difference that emerges between the original play and its revision is that the characters have been renamed. Still named after rivers or bodies of water (with the exception of Pipeline), now Obida has become Imo, Atlantic Oceana, and so on. In Chapter 4 on documentary films, we encountered the nonhuman as a category-confusing horror story. In the last chapter, on Nollywood films, we saw an equivalent conflation of the non/human through the corporation assuming the status of personhood. Here, the nonhuman retains a measure of distanciation since it is never possible to fully inhabit or identify with the radical other.[6] Even the names, of course, are cultural designations for natural phenomena. The plays' dialogue gives voice to the nonhuman, conveying the complex entanglement of human, landscape features, and other species. As the women wait at the fuel pump, knowing full well that they are victims of "artificial scarcity" (37), they reflect on who takes the oil money (25) and who is left "begging . . . just able to survive" (26). Speaking both for and as the nonhuman, they assert their interdependence: "And look around you. See? They're not even killing us alone. The trees too!"—they go on: "They've killed everything with their oil pollution and spillage. We cannot breathe clean air. Fishes die or get fried in the polluted simmering rivers. Water—water everywhere. But we have no clean water to drink! And now we lose the land too?" (27) In their shared experience of catastrophe—however differently felt—the human and the nonhuman double each other without collapsing into each other.

In one of the odder twists in Onwueme's play(s), the Chief and the Government Official decide to set up a "day-care nursery for [the] pets and puppies" (157) of the oil director and others. We can dismiss this turn of events as melodrama (like Nesbith does in his review [256]), or consider it the absurd, though logical, endpoint of devaluing the human: "CHORUS: Abomination! Abomination! Abomination! They oppress us with their pet animals. OBIDA: They hate us and love their beloved beasts!" (158–59). The women invert the moral revulsion of this abomination with the threat to make "Delicious 404" (160), a dog meat delicacy, which triggers physical revulsion in the director (something he notably does not feel about the various atrocities inflicted on the people around him) (166). Just like the men fear that the women in Clark's play have turned into goats into order to inflict their excremental revenge, so too Onwueme's women are suspected of turning into dogs to escape prison (182). Such magical transformations endow the women with power (specifically the feminine power of witchcraft) but also place them in tension with the (masculine) power of the law, which can decree that goats be banished or dogs be shot.

---

[6] Jack Davis dissects Brecht's various stances on nature and the nonhuman in "Two Against Nature? Brecht, Morton and Contradiction," pointing out that, among other things, "Brecht's dramatic theory aims at disclosing the ideological dimension of supposedly 'natural' relations" (217). Reading Davis underscores petroleum's deep penetration into Brecht's thinking, as it frequently arises to suggest "an omnipresent, inescapable web of interrelations between nature and culture" (223).

In an explanatory note about her cast of characters' watery names, Onwueme explains that they are *"like the swift currents of water that defy limiting boundaries (or 'arrests'!)"* (1). Throughout the play, the law and its agents command that people be arrested, a theme that escalates when we hear that, in addition to the kidnappings, the pipeline is on fire. Officials pledge to "arrest the situation" (174), but the police respond, "Saar. Already, the whole nation . . . I mean village is under arrest. Old, young . . . everybody" (184). In the epilogue, "A Nation in Custody," leaders convene a trial to "show the world that people cannot take the law into their hands" (186), but really it is the nation and its leadership rather than the women and youth who are on trial (Eke 11). The courtroom scene ends when the leaders come face to face with the consequences of their actions and the women demand dignity (220). Their chorus—"And so she said it!/She said it-Said it-What Mama Said" (221)—resists arrest. It cannot be stopped as it spills over, out of the reach of the law, off the stage, into the audience, and into the next iteration of the play.

Fixing Things

If we needed any more proof that everything—and especially everything Nigerian—reduces to petromodernity, the dog meat dish 404 got its name from the Peugeot vehicle with the same number. In it, every part of the dog's anatomy corresponds to a car part: "This conscript was explaining the cuts of dog meat. The whole animal, he said that's called Peugeot 504. Or wait, wrong model, it was 404. So it follows, the head, that's the gearbox. To order a leg, you ask for the tyres. The head was headlights, hang on, no, that was the gearbox" (Adetunji 45). Lydia Adetunji gives us this unsavory breakdown in her 2011 play *Fixer*, which draws a set of characters together around the *wahala*, or trouble (Adetunji 17), brewing about an oil pipeline that has been built across the Sahara.

Originally conceived of in the 1970s, the Trans-Saharan pipeline is a real, if incomplete, thing. A gas (not oil) pipeline,[7] the idea had been revived a couple of years before Adetunji wrote her play, when Nigeria, Niger, and Algeria agreed to proceed with the joint venture. Although it was scheduled to open in 2015, no progress was made on it between 2009 and 2019 (Sofiullahi). Since 2019, only a section of the Nigerian stretch, the Ajaokuta–Kaduna–Kano Natural Gas Pipeline (AKKP), has been built, and the future of the rest looks uncertain due to political instability in the region, a worldwide gas glut, and pipeline

---

[7] In addition to its oil abundance, "Nigeria holds Africa's biggest and one of the world's biggest gas reserves" ("Ajaokuta–Kaduna–Kano").

saboteurs (Sofiullahi). Despite its grand ambition, the AKKP will end, unremarkably, at a gas station at Kano ("Ajaokuta–Kaduna–Kano").

For the most part (other than the romance output of the Kano collective), and for good reason, the majority of this book has concentrated on art forms produced in the southern part of Nigeria, where the Delta's rich resources coalesce. *Fixer* is set in the North, although the pipeline carries a familiar set of political and social woes along with its fossil fuel. Characters complain of petrol queues (Adetunji 14, 58) and once again associate oil, excrement, and odor: "What am I supposed to do? When everything is stinking like shit, you think only me I will smell like perfume? . . . The smell of oil dey draw you" (67). The plot revolves around a fixer, Chuks, arranging to bring two British journalists, Dave and Laurence, to interview militants who are suspected of sabotaging the pipeline, also something we have seen before, as in Andrew Berends's *Delta Boys*. By deliberately blurring the two energy resources, gas and oil, and by using the infrastructure of the pipeline to link the North and South of the country, Adetunji resituates the petropolitics of the Delta into a national—and international framework. In doing so, she goes against the grain of much Nigerian writing about oil, which tends, as Oyeniyi Okunoye observes in an essay on Niger Delta poetry, to assert an identity localized in the oil-producing region (430) via a "symbolic severance" (421) from the rest of the nation. This divide, Okunoye reminds us, derives initially from the arbitrary "action [of] the British colonial establishment to forge a nation out of the two protectorates that existed in what came to be known as Nigeria in 1914," and persists in what, "in contemporary Nigerian political discourse, has come to be known as the national question, a euphemism for a call for the interrogation of the Nigerian project" (426). In the play, Laurence, assuming Chuks is Northern, accuses him of being indifferent to the threat the pipeline poses:

> LAURENCE. There are people down South been fighting the oil companies
>   for decades. You know whats going on? You want them building pipe-
>   lines through here now? Doesn't it bother you? As a northerner?
> CHUKS. I'm not from the north.
> LAURENCE. You're not?
> CHUKS. You cannot tell from my name? (Adetunji 52)

Chuks, short for Chukwuma (Adetunji 36), is an Igbo name, but Laurence, a Black-British person, fails to appreciate the identity politics it carries. In that oversight, Laurence assumes a shared national interest in thwarting the unnamed "consortium" that has built the pipeline. After a pause, Chuks philosophizes, "North, South, it is how it is" (52), dismissing both the distinction between the two and his own agency or attachment to any location.

The British-born, Nigerian-raised Adetunji began her career as a journalist. A *Guardian* story on new political playwrights in 2010 quotes her on the switch:

> "I'm still a believer in what journalism can do . . . but it's interesting for me what works better as journalism and what works better as theatre. I feel like drama can sometimes get to the nub of things better, there's more freedom and no requirement to be balanced." Also, she says, she loves the excitement of "having people and live human emotion in front of you. It's much more immediate." (Edgar)

Fixer explores the tension between journalism and drama by performing reporting; beyond the fact that, throughout the play, characters discuss the business and politics of journalism, we frequently watch them dictating their stories as part of the drama. Uttering text changes its nature, from written to spoken, and plays with form by integrating reportage into the script.

The first time this happens, Laurence is waiting while Dave interviews a militant. Laurence *"gets out his dictaphone"* in order to capture some "color" for his story (50). He dictates scene-setting paragraphs, each of which prompts a correction from Chuks, like his description of the pipeline as a "gleaming iron snake," to which Chuks replies, "It does not resemble a snake. . . . Also, the material is aluminium" (51). Later, in the last scene, Dave and Laurence discuss whose story "got a better show" and agree that Laurence's "Pipeline in Peril" was "Crunchier. Snappier. More . . . colorful" (75–76). Never mind that they could just as easily be describing a candy bar. When Laurence files his report, he dictates it over the phone to his copytaker, complete with punctuation: "LAURENCE (*dictating slowly and clearly*). Last night the situation appeared to be spinning out of control as militants vowed to step up their campaign. Point, par." (63).

Within the structure of the play, this is a performative utterance in that the speech act makes something happen, a thing (the newspaper story) that happens to be writing, so that what is being uttered is simultaneously oral and textual, both in terms of the dialogue and the written script. It also intends to make change happen, to "get the local people's story out" and "Give Africa a voice" (38). But what happens if the utterance does not change anything? The question of whether they have done anything or nothing troubles the journalists (and the audience). On the airplane back to London from Lagos, Laurence tries to convince Dave that his story had impact: "LAURENCE. It was a good story. Got the consortium on the back foot, got people to look closer at the pipeline, at the boys. It changed the state of play. DAVE. Nothing changes" (77). It seems that the consortium might have paid off "the boys," resulting in a stand-down on both sides. If so, it doesn't matter whether anyone read their stories, and, perhaps, no one did (77–78), leaving only us to hear them. The politics of the performance within the play vacates it of any discursive or actual power—the only thing they did make happen was probably to have caused Chuks' death (and therefore doomed his daughter). The play ends with them absolving themselves even of that; Laurence speaks the last line: "These things happen" (79).

On stage, actors perform the performative utterance. Their motives are doubled-edged, however: The play wants to both critique the knee-jerk, self-interested instincts of its characters and it wants to do some of the "work," via drama, that journalism tries to do. Although it is deeply cynical, sardonically showing us how everyone is using everyone else, it also has an earnest streak. Perhaps it can remind us that "energy is not a simple climate change issue but a tangled mess of geopolitical obligations" (Freestone and O'Hare 16) or help us figure out how to write about Africa, even if it's a truism that there is "no call for Africa stories at the moment" (Adetunji 26). Listed in *100 Plays to Save the World*, *Fixer* sets out to fix things; it hopes that drama will be the fix much in the way that collection's editors wish: "Theatre must imagine the future, and help us reach toward the bold, humane, quick thinking we are going to need" (Freestone and O'Hare xiv).

The politics of performance for the theatrical work of art itself are thus also vexed. *Fixer*'s satire of selling Africa is straight out of Binyavanga Wainaina's "How to Write About Africa." "Your hero is you (if reportage)" (93), writes Wainaina, "Whichever angle you take, be sure to leave the strong impression that without your intervention . . . Africa is doomed" (92–93). Use "broad brushstrokes" but make sure to include "(unspeakable) suffering" (93). In a world where everything is for sale and everyone is a sell-out, Chuks knows that "anything you can see, somebody will buy it. Is suffering not a commodity? Bad news? Pack am, sell am" (Adetunji 15). This is why the journalists are really there: "They are coming to Africa looking for bad things, horror film. I can give them the cinema" (Adetunji 68). We have seen the ability of reporting to turn documentation into horror film before, and the question of who controls what we see is at the heart of *Fixer*. The reason consultant Sara is "up north" is because she is also a fixer, of a sort: She helps clients—like the consortium—with "image problems . . . offer[ing] a different narrative" (Adetunji 4). Doing this, she says, is "an art" (Adetunji 4), a line that invites comparison to the playwright and to the messaging potential of the dramatic arts. For Adetunji's satire to work, the play must succeed in making its audience think or do something differently after watching it. Otherwise, it merely repeats the performance of reporting Africa, which is already its own parody.

In their introduction to *100 Plays to Save the World*, Elizabeth Freestone and Jeanie O'Hare reflect on the works they have collected in that volume:

> The wilder plays are born from the imaginations of writers whose neighbourhoods are burning and whose homes are flooding. The further you are from the daily lived reality of the climate crisis, the quieter and more formally conservative the plays. This collection has revealed how the world is reshaping itself violently in the physical realm and how that is impacting on the reshaping of stories we need to tell, not just for now but for generations to come. (xiii)

Indeed, *Fixer* is a much tidier, tighter play than the "wilder" ones we've previously discussed. It risks, however, a kind of dramatic solipsism if it all comes down to the final pages in which Dave and Laurence dish out liability for what happened, actuarially attributing 80 percent of the blame to Chuks himself (79). If we are left feeling as they do, that it was "for nothing" (78), then the play is a closed circuit, just a performance, a repeatable ritual, (theoretically) endlessly iterative without introducing change.

Adetunji tries to change the narrative through a slow reveal about how the political is personal. When we first meet Chuks, he advertises his services with a portfolio of tragic snapshots—a victim of battery acid, a woman beaten during a religious riot, a five-year-old girl with polio. Although Chuks tries to push the polio angle, Dave is only interested in the pipeline (8–10). Gradually we realize that the girl whose "legs don' bend make K" (69) is Chuks' daughter. If the pipeline fire is the "better show," the bright candy wrapper of a story, at the soft center of the play is a nougat of personal pathos. It's the thing that humanizes Chuks in a way that the other characters don't merit; it defies Wainaina's direction to "avoid having the African characters laugh, or struggle to educate their kids, or just make do in mundane circumstances. . . . African characters should be colourful, exotic, larger than life—but empty inside, with no dialogue, no conflicts or resolutions in their stories, no depth or quirks to confuse the cause." On the other hand, it breaks Wainaina's taboo against "mention of school-going children who are not suffering from yaws or Ebola fever or female genital mutilation" (92). Freestone and O'Hare notes that "the reoccurrence of [polio] is still keenly felt by many Nigerian families"[8] (16), and of course the disease is a family matter, but it is also, like oil, a matter of geopolitics (Jennifer Cole). The personal is political.

The geopolitics of oil in *Fixer* entangle players local and global, north and south, advantaged and disadvantaged. In this way, it is what Jennifer Wenzel understands literature in general to be: "a mesh of relations in which the liberatory and immiserating implications of globalizations—old and new—are knit" (*Disposition of Nature* 8). The unspecified international consortium in the play disrupts local conditions even as it replicates that disruption worldwide: "You know what the consortium does, don't you?" Laurence presses Sara, "Moves in, starts siphoning the oil away. Not for the locals' benefit, all for its own interests. They've made a mess down south" (20). That mess may be specific to that locale, but it is just one point in a greater messy geography of petroleum: "Lagos isn't so bad," Sara acknowledges, "I mean, there are better oil postings. Aleppo's nice. But Baku, after Baku, Lagos is great . . ." (22). They are "part of something bigger" (23), an oily marriage of private and public sectors, at home and abroad. (Closer to home, a Minister of Finance throws lavish parties, presumably having siphoned off some oil money for himself [5].) Originally performed as part of the London via Lagos festival, the play brings together British, American, and Nigerian characters by way of a series of nonnational spaces such as hotel lobbies and airplanes. Even in the scene that takes place in the militants' camp the set consists of "*Two airline seats, side by side. Not on a plane, but somewhere outside*" (45), a reminder of the transnational, transactional nature of the material negotiations around the possession and distribution of the resource that are mirrored in the discursive negotiations over who gets to tell the story of oil.

All negotiations accrue around the pipeline, a piece of infrastructure that structures the play while remaining out of sight. Despite its status as a fraught and constant reference point, we never see it. It is, however, visible from space (Adetunji 3). As an unseen but also hypervisible hyperobject, it breaks up the "visibility/invisibility dyad" Wenzel describes: "Among the things concealed by the visibility/invisibility dyad are the subtle interplay of invisibility and hypervisibility. Some things that seem invisible are actually hiding in plain sight (or even subject to surveillance); other things that seem spectacularly hypervisible remain for all practical (and political) purposes unregarded and unapprehended" (*Disposition of Nature* 14). Unlike the invisible infrastructure we take for granted—because it is actively invisibilized (Larkin, "Promising Forms" 186)—the pipeline cannot

---

[8] We have had a vaccine for the highly contagious disease for approximately three quarters of a century. By 2018, the UN had declared that polio was no longer endemic in Nigeria, but at the time of writing, efforts continue to try to eradicate polio in Northern Nigeria (UN News).

be fully invisible. First, it is incomplete: "Infrastructural invisibility corresponds to an ideal which is only ever tenuously achieved. In those places where infrastructure seems always in need of repair, or where it is unfinished or disappointing, infrastructural invisibility is an elusive goal" (Hetherington 42). And second, because "material infrastructures, including . . . oil pipelines . . . as dense, social, material, aesthetic, and political formations . . . [reveal] fragile and often violent relations between people, things, and the institutions that govern or provision them" (Appel, Anand, and Gupta 3). The consortium's pipeline is a nexus of these violent relations, drawing to it the locals who want to blow it up (3), the army sent in to crack down on them (63), and everyone else who wants to fix the situation.

Although theorists of the "infrastructural turn" tend to champion the way that "communities worldwide [that] face ongoing problems of service delivery, ruination, and abandonment . . . use infrastructure as a site both to make and contest political drama," and thus make themselves visible as demanding subjects (Boyer 223, 3, 23), the negative lesson of Adetunji's play is that such resistance is often reabsorbed into the very institutions, bodies, and corporations of which they are making demands. We learn that the men have been tapping the pipeline and selling the oil on the black market (Adetunji 70), effectively financing the movement against oil with oil money. Moreover, we suspect that the "financial accommodation" the militants and the consortium came to that "managed the situation" involves employing "the boys who are blowing up the pipeline to provide security to prevent themselves from blowing up the pipeline" because, who better? "They have considerable experience with the pipeline. They know it well. Best candidates for the job" (Adetunji 8, 3, 58). These are the ways in which the system allows for spillage and waste (material or not)—by building containment into infrastructure.

Fixer neatly contains its action. It comes full circle, incoming on the plane to Lagos, outgoing on the flight to London. Inside that dramatic structure lives plenty of irony and satire, exposing the gaps between what is and what should be. The editors of *The Promise of Infrastructure* call "infrastructures . . . critical sites for the distribution of life and a key locus for the performance (and theorization) of politics and polities today" (Anand, Gupta, and Appel 21). Adetunji takes that at face value, giving us the performance of infrastructure politics as petrodrama. Although all the plays read here are "structured and conditioned by the oily machinations and social relations fuelling the extractive politics of the world's most inflammatory resource" ("Commentary" 1), per Graeme Macdonald's definition of petrodrama, *Fixer* is dramatically different from the wilder others, with their "excrement, excess, [and] superfluity" (Esty 54). *Fixer* narratively and formally refuses excess. It does not spill over.

Queering Oil

Oil enables and evinces all manner of gender relations, and in doing so it changes them: It heightens displays of militant masculinity; exacerbates traditional feminine gender roles, such as women as "fixers" of both public and private dramas; and compromises sexual and marital relations. In other words, oil queers gender relations in the sense that *to queer* means to change or to act in a way that disrupts normativities (Lindsay)—think of how conventional love and marriage get warped into adultery and incest in *The Oily Marriage*, for example. At the same time, if all gender is fundamentally performative (à la Judith Butler), petrodrama exaggerates that performative aspect, often bunkering characters into narrowly defined roles and reducing them to types. Sometimes, petrofiction is overtly queer—certainly we have to read Binebai's "interest in the anus" through the lens of queerness and not only as "symptomatic of the general desperation and waste of the region as a result of petro-extraction" (Ajumeze 110).

Oil queers things because oil is queer. We can build on Rebekah Sheldon,[9] who builds on Karen Barad, to reach this conclusion. Sheldon, writing about the figure of the child as a technology to save the future (25), contends that environmentalism relies on stillness and stasis to imagine a discrete, known future. "It is against this reproduction of fixity that I seek to situate the queerness of matter" (Sheldon 30), she writes. Barad sets out to rethink queerness and performativity together, beyond the human and the humanistic, talking us through amoebas, atoms, and more to arrive at the notion of "nature's queer performativity," a claim "that there is something inherently queer about the nature of matter" (29, 39). Although Barad does not discuss oil, she leads with the example of a news story about amoebas "oozing through Texas soil" which she says speaks to the issues in her paper, many of which apply here, too: "concerns over foundations, stability and instability, reconfigurations and shape shiftings, nonhuman agency, queer critter behaviors, fear and moralism, and nature/culture, micro/macro, temporal, and spatial boundary crossings" (25). The language she calls our attention to— of a "vast and sticky empire . . . oozing along in the muck" (Yoon qtd Barad 25) could easily be describing our matter at hand. It also sums up the queerness of matter—its "indeterminacy" (Barad 39), how it "is all iteration and permutation" (Sheldon 30). Oil—mutable, movable—must then be the queerest of matters, changing its own form and reforming the forms it inhabits.

---

[9] In a footnote, Sheldon shares a handy bibliography of books that situate themselves at the intersection of queer theory and ecocriticism (200n14).

CONCLUSION

"'It's not a joke,'" says Dandy, one of Ken Saro-Wiwa's characters from his *Basi* stories (121). That's a laugh line if you're a viewer or reader of the series, because, of course, it's part of an elaborate joke that you are in on. Before his novel *Sozaboy* (1985), Saro-Wiwa wrote Nigeria's most popular sitcom, *Basi & Co* (1986–90). "30 Million Nigerians Are Laughing at Themselves," read a 1987 *New York Times* headline about the show that built up a common Nigerian national identity by creating a shared sense of humor (Brooke). James Hodapp's essay, "A serious television trickster: Ken Saro-Wiwa's political and artistic legacy in *Basi and Company*," is one of only a few scholarly examinations of *Basi*. Hodapp takes the sitcom seriously, noting that while it is often thought of as marginal to Saro-Wiwa's political project, it is itself a political commentary, addressing serious topics like police bribery, corporate embezzlement, capitalist greed, and the inability of the government to provide the most basic infrastructure (505). Said Saro-Wiwa in an interview, "I have used it to excoriate Nigerian society at the moment" (Brooke).

With *Basi*, Saro-Wiwa not only wields words as weapons by exploring the scathing side of satire, he plays with form. Hodapp explains how he does this within the sitcom format: referencing the ways in which it shares aspects of traditional storytelling (episodic structure, character types) and reinventing it as a topical, didactic, nation-building tool. Even as it pricked at Nigerian politics and power structures, it moralized about (the lack of) a work ethic and "proper English" (Saro-Wiwa qtd Brooke) (ironic, considering the "rotten English" of *Sozaboy*). In the book version of some of the sitcom episodes, *Basi and Company: A Modern African Folktale*, we see Saro-Wiwa's deliberate interplay between forms. His author's note sketches a reverse lineage which frames an older narrative form within a newer one, so that folktale is proto-sitcom:

As an adult, it has dawned on me that the form in which we heard the exploits of the Tortoise is very much of the genre of the television series.

> Consequently, in transforming the folktale into a contemporary idiom,
> I adopted the format of the television comedy series, and it is no wonder
> that *Basi And Company* first appeared as such.
>
> In presenting it in book form, I have maintained this format. (Saro-
> Wiwa, *Basi*, author's note)

In her essay on the Basi book, Jane Wilkinson calls this movement between forms a "series of re-elaborations" (Wilkinson 255). Wilkinson's idea of the "second-new" (a phrase she borrows from *Basi*) describes the recycled, renewable nature of the Basi texts, a quality she sees as both "ceaselessly self-proliferating" and "a pattern that can be introduced into new circuits of meaning, providing the 'second-new' building-blocks for new constructions" (264). The characters are "replaceable, recastable" (265), the plots recirculated, and political power is concentrated in anonymous, generic subjects, but these are also the reasons why "the rule of replication, substitution and exchange makes it—theoretically—possible to place other figures in the subject slots, changing the combination to fit the situation, the actors to fit the part, the words to fit the phrasing" (266).

Basi's catchphrase is, "To be a millionaire, think like a millionaire!," and at the heart of every Basi escapade is a money-making scheme. So much of Basi repeats, including—especially—the economics of "dealing as an end in itself," where "what remains of the physical 'market-place' in the actions of the protagonists is the exchange and circulation of immaterial words and figures. The commodities and activities they refer to have been effaced, delivery indefinitely deferred. What is negotiated is not even a 'representative' value but a purely illusory, fictitious one" (Wilkinson 260). Taking this further, Jacqueline Bardolph asserts that Basi shows the whole system to be fictive, as expressed in a language of empty words and euphemisms repeated like magic formulas (19–20). She gives "commodity" as an example of a word "taken up and bounced about between Basi and his friends," disconnected from any real goods (19). Oddly, though, none of the Basi scholarship considers oil in this light, despite the fact that not only is oil the ultimate in speculative, irrational commodities with fictitious value,[1] but it is also a touchstone reference for Basi—and one of the many things he tries to trade in.

---

[1] For more on this idea, see Mazen Labban's essay "Oil Spill: Inside the Global Market for Crude" which describes the "irrational system" that produces the price of oil as an abstraction resulting from hypotheticals.

Oil surfaces in back-to-back stories in the book *Basi and Company*. In "The Bank Loan," Basi and friends, hoping as always for an "easy way to the millions" (108), try to get a collateral-free loan in order to become "gentlemen farmers" because "oil was a wasting asset; money in agriculture would make the country prosperous for all time" (105). The idea that oil will decline in value over time might come closest to Basi actually perceiving worth for a change; certainly for poor Nigerians like the residents of Adetola Street there is no return on oil futures. A character sums up their sense of economic exclusion in the following story, "Countertrade": "All I hear about is the oil boom. But where is it? Boom! boom! boom! but not around the 'Palace' here" (119). The oil boom is always everywhere but here, its elusiveness dictating the need for redoing the scheme and repeating the plot as the characters try again to capture its slippery wealth.

In "Countertrade," the get-rich-quick plans pile up on each other, each trade adding to the pyramid scheme. Josco, one of the "company," wants to sell bunker fuel to a captain in port but doesn't have the money to buy it on the black market, so he borrows it from Basi who borrows it from Madam the Madam, the landlady. They give the captain what looks like fuel—but it's only a thin layer floating on top of water. He, in turn, trades them what looks like brandy but is actually colored water. To stall Basi, Josco tries to palm off stacks of "special dollars" (131) (blanks hidden under a few real bills). In the end, with exquisite predictability, the entire pyramid collapses and the narrator (another young woman leaving to get an education) delivers the moral of the story: "I had begun to see how greed and selfishness could together make fiction of life" (134). Each of these false commodities is equally fungible and worthless. None of the doctored liquids purchases the characters any kind of liquidity— economic or otherwise. The oil is unusable; the brandy undrinkable, although in the real economy of "the unofficial market," Josco and friends could easily have bought bunker fuel from the men who undercut their bosses by siphoning it off and selling it (129–30).

Writing about *Basi* the sitcom for the *New York Times*, James Brooke draws a direct connection between Nigeria's oil wealth—the country having "won the oil lottery"—and Basi's "get-rich-quick mentality." He quotes the show's director, Uzorma Onungwa, on one of the moral messages the sitcom was intended to convey: "'There is this tendency of people to want to make millions without really working for it,' Mr. Onungwa said. 'We are trying to teach Nigerians that working hard really pays—not lazing around.'" Hodapp expands on Brooke, connecting the sitcom's recurring theme of greed to Saro-Wiwa's larger political project. Talking about the episode in which Basi mocks up a literal money-making machine, he says it refers to "exploitation in a way that would later become a hallmark of Saro-Wiwa's argument against Shell and the Babangida and Abacha regimes. The Ogoni lived and worked on their own land, but did not receive the billions made from it" (Hodapp 510). He goes on: "The entire ethos of the nation's economic trajectory is based on a Basi-like scheme reminiscent of the machine. Greed is not only fuelled by the universal selfishness of humanity, but by the specificity of the oil boom in Nigeria and the immediate aftermath" (510). It is the empty promise of oil money that creates the conditions for Basi's attitude, and the pyramid scheme that structures the "Countertrade" story reflects the pyramid scheme of the oil industry—extraction without recompense, resources and profit shunted off from their rightful owners, irrational pricing unhinged from actual use-value.

Hodapp views "The Machine" episode through the lens of the oil boom:

> The boom, and post-boom ethic, makes an empty box with a sheet over it a fitting metonym for the period because it is able to create money, at least temporarily, out of nothing. The empty space at the centre of the box does not even attempt to hide the nature of the box and its workings. Rather, a thin sheet and the promise of instant wealth creation, and ice cream, do not sound unreasonable because the core of the boom era was equally void of content, yet produced absurd amounts of wealth for a select few. (510–11)

Beyond oil, the empty box reads as a signifier of the machinations of a bigger false economy and the empty promises of capitalism. There is nothing at the base of this system; just a belief in an alchemy through which money becomes more money by subordinating living labor to a commodity fetish.

"The easy availability of oil money" (*Basi* 119) somehow never comes to Basi, but his eternal optimism persists. In every story and every episode, he and his company try, usually through devious means and magical thinking, to recover what they feel is rightfully theirs, or at least what they feel promised by the perception that "All over Lagos people are making it real big on oil" (*Basi* 119). The many references in the book (at least half a dozen) to recovering property speak to this longing. The "hope and belief" Saro-Wiwa places in his trickster (Bardolph 22) allow for a positive spin on his otherwise repetitive words and actions—they make us fond of and amused by him, not frustrated and disheartened. They signal a potential reclamation of history, agency, and resources made possible by the "hybridity, rupture and supplementarity" (Wilkinson 258) of the postcolonial project. Commenting on the details of the stories' settings and characters but gesturing toward a larger contention, Wilkinson says, "The erasure of the past makes room for its creative recovery, for an elastic re-invention of origins" (258). Basi shows the way to this creative recovery. Greed and foolishness might make fiction of life, exposing the "dream world" in which everything is "as worthless as pieces of fiction" (Bardolph 20), but by the same token, Basi is the author of his own fictions.

Perhaps it is foolish to follow the fool. After all, we have received warnings about the latent conservative politics of recovery. Stuart Hall complicates the idea that "the subjects of the local, of the margin, can only come into representation by, as it were, recovering their own hidden histories" with the caveat that those returns, those rediscoveries of identity, can also constitute a form of fundamentalism (183–84). Roderick McGillis's "Postcolonialism and Recovery" also treats the concept gingerly:

> Recovery means covering things with that which has served as covering for generations; recovery is not so much renewal as it is return. Recovery seeks to return us to the condition we feel most suits us, that is, the condition of supremacy. Recovery in this sense is deeply conservative, and in a colonial sense it perpetrates the hierarchy of colonizer/colonized. Recovery also means recuperation. In this sense, recovery is the process of healing, and returning to a condition of health before some alien germ—say, a colonizer—had initiated a period of ill health. In a postcolonial sense, recovery of this sort is the reestablishment of an independent identity, rather than one fashioned by the language and perspective of the colonizer. Recovery of this sort is, however, beyond reach, in the sense of returning to the condition prior to the coming of the colonists. Recovery must mean recreation, the fashioning of something new. (22)

In short order, McGillis ends up somewhere very different from where he began: Recovery is not renewal, it is return, he says. But return is impossible, he says, and so recovery must instead mean something new.

Recovery as newness directs us imaginatively forward. In the context of this study, recovery helps us think about the crises that petroculture throws up and the ways in which different petroforms meet them. Foolish or not, I propose mobilizing recovery as generator of counter-futures (301), to borrow a phrase from Kodwo Eshun's experiment in Afrofuturistic theory in which he speculates that "a team of African archaeologists from the future . . . would be struck by how much Afrodiasporic subjectivity in the twentieth century constituted itself through the cultural project of recovery" (287). The *Oxford English Dictionary*'s entry on *recovery* offers us a script by which to practice this cultural project. What follows is a formal experiment in reading through the

OED's senses of the word *recovery* to rehearse other possible outcomes of and remediations to the status quo:

Sense I.i.2.a. The fact or process of gaining or regaining possession of or a right to property, compensation, etc., by a legal process or judgment.

Legal remedies against Big Oil have been pursed domestically and abroad for decades, and while the law can be an effective means of resistance, justice for the Niger Delta has proved to be "slippery" (an adjective evocative of oil's feel) (Akinbobola). For instance, in 2024 the Nigerian Supreme Court agreed to hear Shell's appeal to a 2020 High Court ruling that had ordered it to pay $878 million to communities in Rivers State (Eboh).

Sense I.i.3.a. The restoration of a person (or more rarely, a thing) to a healthy or normal condition, or to consciousness.

At least since Ken Saro-Wiwa helped draft the Ogoni Bill of Rights in 1990, which has been called a "landmark document in indigenous struggle" ("Ogoni Bill"), there has been a consciousness-raising campaign around local rights to resources in the area. A more recent, and more fraught, take on this history is Omolade Adunbi's theory of "oil consciousness," which argues that the creation of an oil consciousness in a population can mobilize dissent on the one hand, but on the other, it also means that "in the Niger Delta, people must define themselves in relation to state and legal entities that deny them access to what they claim as their own" (Adunbi 17).

Sense I.i.3.b. The cure or healing of an illness, wound, etc. Obsolete.

The petropoetry of Tanure Ojaide, in which the Delta has been wounded by the trauma of oil extraction, speaks to this definition directly. Oil is the lifeblood that flows from "an immeasurable wound":

> Barrels of alchemical draughts flow
> from this hurt . . .
> The inheritance I sat on for centuries
> now crushes my body and soul. (*Delta Blues* 21)

Sense I.i.4.a. The regaining or restoration to one's control or possession of a thing lost, stolen, or otherwise taken away, retrieval; the possibility of recovering such a thing.

Practically, see the many instances of bunkering, sabotage, and so-called theft we have encountered in this book. Conceptually, see the sentiment that oil is the Delta's stolen birthright (Orr).

Sense I.i.5.b. The reclamation of wasteland, esp. land previously under water, for cultivation or construction.

The idea that the Delta has become a "wasteland" sits at the center of the popular imagination about oil production in the region (to the extent that anyone outside the Delta has such an imagination). As a case in point, Said Adejumobi's 2004 opinion piece, "The Niger Delta: An Open Sore of A Shameless Country," does not mince words in this regard: "The Delta is a wasteland, a site of griming poverty, disillusionment, despair, frustration, and hopelessness." In contrast, projects like the National Mangrove Restoration Project (Chijoke) sanguinely contribute to the slow rehabilitation of the Niger Delta.

Sense I.i.5.c. Economics. The process or fact of returning to an improved economic or financial condition, esp. on the part of a country or its economy.

Visions for a more equitable economy range from reparations for local communities to a future in which energy poverty in the Global South is addressed via a just transition to renewables. Many of the texts read in *Petroforms* call out financial inequities, from the individual to the collective. The wives revolt in Clark's play because they reject the claim that, in the matter of "the money sent by the oil company operating on our land . . . a most fair and equitable settlement you will never find in any other society, near or far" (1).

Sense I.ii.8. The extraction of a useful substance from waste material. . . . Also: the extraction of a usable substance from a mixture or from raw material.

See developing technologies in remediating toxic waste, the reuse of waste as a resource, and the recovery of crude oil from petroleum sludge, including various methods of bioremediation that have been used in Nigeria to reclaim polluted soil (Ayotamuno et al.).

Sense I.ii.9.a. Astronomy. The observation of an astronomical object following an extended period during which it has not been visible or observed, esp. the sighting of a periodic comet.

The planetary trend in postcolonial criticism makes visible degrees of resource depletion, our mutual entanglements and interdependence, and our radical difference from the nonhuman. Amitav Ghosh's "tiny planet" (*Nutmeg's Curse* 10) of a nutmeg lets us see planetary crises. Among these is our climate crisis. As Jennifer Wenzel explains, gas flaring in Nigeria contributes to global warming on a planetary scale (*Disposition* 92), and there is a growing public interest in legally trying fossil fuel companies for homicide for climate-related deaths (Noor).

Sense I.ii.11. Chiefly British. The retrieval and transportation of a damaged or disabled vehicle from the site of an accident, breakdown, etc., to a place with facilities to carry out repairs, or to a particular destination.

Regard the *Battle Bus* as an assemblage of discarded parts recovered and reassembled in the service of the petroform, as well as efforts to retrieve the *Battle Bus* from government confiscation.

Sense II.12. The possibility or means of recovering, or being restored, to a former, usual, or correct state; help or remedy. Chiefly in negative construction; now only in beyond (also past) recovery.

See, instead, the remediative value of the fictive, the imaginative, and the artistic.

Sense II.14.a. Restoration or return to health from illness, an injury, etc.; an instance of this. Also used with respect to an injury: restoration to a healthy state. Also with a modifying word, as full recovery, complete recovery, etc.

Brutalizing extractive practices leave both human and nonhuman agents in a chronically unhealthy state. In 2011, the United Nations predicted that systemic contamination in the Delta would take thirty years to rectify (Vidal).

Sense II.15. Materials Science. Reversion or return of a material, object, or property to a former condition following removal of an applied stress or other influence.

Stressed by their encounters with petroleum, aesthetic and other forms do not merely revert (and to wish otherwise would be "deeply conservative," as McGillis warns). Instead, "to what extent bodies possess the property of total recovery of form, when relieved from a strain, is still a matter of doubt" (Mahan qtd "recovery, noun").

The petroforms discussed here engage with the project of recovery by attempting to regain control of what has otherwise been taken away (both the resource and control over it). The possibility of recovering such a thing remains fragile, as do the forms in which it collects. We have seen how oil changes things, sometimes bewilderingly, as for Basi: "I did not quite understand how the presence of oil in the swamps enabled houses to spring up fast on Adetola Street. But whether I understood or not, the houses went up" (Saro-Wiwa, *Basi* 11). Petroleum, extracted from the Delta, subsequently diffuses into magical oil money that alters places and their infrastructures from a remove. It shrinks and expands the dimensions of these spaces. Adetola Street represents the limits of Basi's world but is also a cosmopolitan marketplace: "The shops were littered with all sorts of goods from different parts of the world, thanks to the easy availability of oil money" (Saro-Wiwa, *Basi* 119). Within the confines of Saro-Wiwa's textual infrastructure, Basi endlessly repeats himself, trying but unable to recover his fair share. But forms and their infrastructures are not stable—for Saro-Wiwa, they are sometimes short story, sometimes sitcom episode, sometimes play, sometimes children's story (Wilkinson 255). Elsewhere, they are sometimes feature film, sometimes documentary, sometimes horror, sometimes sculpture, and sometimes masquerade. It may be "the work of form to make order" (Levine 3), but petroforms are inherently disorderly. They reflect a politics of creative recovery, a politics that refuses to clean up someone else's spill.

WORKS CITED

Adebanwi, Wale, and Ebenezer Obadare, eds. *Encountering the Nigerian State*. Palgrave McMillan, 2010.

Adejumobi, Said. "The Niger Delta: An Open Sore of A Shameless Country." Edo-Nation.net. 10 Oct. 2004. Web. 5 Apr. 2024. https://edo-nation.net/articles/the-niger-delta-an-open-sore-of-a-shameless-country-1108.

Adetunji, Lydia. *Fixer*. Nick Hern Books, 2011.

Adichie, Chimamanda Ngozi. *Americanah*. Anchor Books, 2013.

Adunbi, Omolade. *Oil Wealth and Insurgency in Nigeria*. Indiana UP, 2015.

The Africa Report, Donu Kogbara, and Belinda Otas. "Black November: Niger Delta Film Spills Powerful Story." *The Africa Report*, 29 Aug. 2012. Web. 17 Jan. 2024. https://www.theafricareport.com/6759/black-november-niger-delta-film-spills-powerful-story/.

Agamben, Giorgio. *Homo Sacer: Sovereign Power and Bare Life*. Stanford UP, 1998.

Aghoghovwia, Philip. "Nigeria." *Fueling Culture: 101 Words for Energy and Environment*. Eds. Imre Szeman, Jennifer Wenzel, and Patricia Yaeger. Fordham UP, 2017, 238–41.

——. "Poetics of Cartography: Globalism and the 'Oil Enclave' in Ibiwari Ikiriko's *Oily Tears of the Delta*." *Social Dynamics: A Journal of African Studies* 43.1 (2017): 32–45.

"Ajaokuta–Kaduna–Kano (AKK) Gas Pipeline." NS Energy. 21 May 2020. Web. 27 Feb 2024. https://www.nsenergybusiness.com/projects/ajaokuta-kaduna-kano-akk-gas-pipeline/.

Ajeluorou, Anote. "Iwowo: Oloibiri Provoked Formal Apology from Gowon to People of the Niger Delta." *The Guardian*. 24 Dec. 2016. Web. 21 Jan. 2024. https://guardian.ng/art/iwowo-oloibiri-provoked-formal-apology-from-gowon-to-people-of-the-niger-delta/.

Ajumeze, Henry Obi. *The Biopolitics of Violence in the Drama of the Niger Delta*. Diss. U of Cape Town, 2018. http://hdl.handle.net/11427/29522.

——. "Petro-Drama in the Niger Delta: Ben Binebai's *My Life in the Burning Creeks* and Oil's 'Refuse of History.'" *Oil Fictions: World Literature and Our Contemporary Petrosphere*. Eds. Stacey Balkan and Swaralipi Nandi. Penn State UP, 2021. 99–113.

"Akeley Hall of African Mammals." American Museum of Natural History. n.d. Web. 8 Apr. 2024. https://www.amnh.org/exhibitions/permanent/african-mammals.

Akinbobola, Yemisi. "Slippery Justice for Victims of Oil Spills: Nigerian Villagers Lose Lawsuit Against an Oil Giant." *Africa Renewal*. Aug. 2013. Web. 6 Apr. 2024. https://www.un.org/africarenewal/magazine/august-2013/slippery-justice-victims-oil-spills.

Akingbe, Niyi. "Speaking Denunciation: Satire as Confrontation Language in Contemporary Nigerian Poetry." *Afrika Focus* 27.1 (2014): 47–67.

Akpan, Udeme. "FEC Approves Contract for Oloibiri Museum and Research Center." *Vanguard*. 8 Feb. 2023. Web. 21 Jan. 2024. https://www.vanguardngr.com/2023/02/fec-approves-contract-for-oloibiri-museum-and-research-center/.

Akpan, Uwen. "Baptizing the Gun." *The New Yorker*. 4 Jan. 2010. Web. 3 Dec. 2012. https://www.newyorker.com/magazine/2010/01/04/baptizing-the-gun.

———. "Luxurious Hearses." *Say You're One of Them*. Little, Brown and Company, 2008: 187–322.

Aldana Reyes, Xavier. *Horror Film and Affect: Towards a Corporeal Model of Viewership*. Taylor & Francis, 2016.

Alkali, Zaynab. *The Virtuous Woman*. Longman, 1987.

"Amniotes." OneZoom. n.d. Web. 16 Nov. 2023. https://www.onezoom.org/life/@Amniota =229560.

Amos, Jonathan. "Munch Inspired by 'Screaming Clouds.'" *BBC News*. 24 Apr. 2017. Web. 5 Oct. 2024. https://www.bbc.com/news/science-environment-39697256.

Anand, Nikhil, Akhil Gupta, and Hannah Appel. *The Promise of Infrastructure*. Duke UP, 2018.

Anderson, John. "Sweet Crude." *Variety*. 15 Aug. 2009.

Angerer, Marie-Luise, Ingrid Richardson, Hannah Schmedes, and Zoë Sofoulis, eds. *Containment: Technologies of Holding, Filtering, Leaking*. Meson Press, 2024.

Appel, Hannah. *The Licit Life of Capitalism: US Oil in Equatorial Guinea*. Duke UP, 2019.

Appel, Hannah, Nikhil Anand, and Akhil Gupta. "Introduction: Temporality, Politics, and the Promise of Infrastructure." *The Promise of Infrastructure*. Duke UP, 2018: 1–38.

Appel, Hannah, Arthur Mason, and Michael Watts, eds. *Subterranean Estates: Life Worlds of Oil and Gas*. Cornell UP, 2015.

Apter, Andrew. *The Pan-African Nation: Oil and the Spectacle of Culture in Nigeria*. U of Chicago P, 2005.

Arendt, Paul. "Say It with Buses." *The Guardian*. 9 Nov. 2006 Web. 29 Jan. 2025. https://www .theguardian.com/artanddesign/2006/nov/09/art.humanrights.

Ashcroft, Bill, Gareth Griffiths, and Helen Tiffin. *Post-Colonial Studies: The Key Concepts*, Taylor & Francis Group, 2013.

Asuni, Judith Burdin. "Blood Oil in the Niger Delta." Special Report. United States Institute of Peace. Aug. 2009. Web. 21 Mar. 2015. https://www.usip.org/publications/2009/08 /blood-oil-niger-delta.

Atta, Sefi. "Independence Day." *Ìrìnkèrindò: A Journal of African Migration* 4 (Jan. 2011). Web. 20 Mar. 2015. https://africamigration.com/issue/jan2011/ATTAH-Independence-Day.pdf.

———. "A Union on Independence Day." *Eclectica Magazine*. Oct./Nov. 2003. Web. 13 July 2015. https://www.eclectica.org/v7n4/atta.html.

Ayotamuno, M.J. et al. "Bio-remediation of a sludge containing hydrocarbons." *Applied Energy*. 84.9 (2007): 936–943.

Balkan, Stacey and Swaralipi Nandi, eds., *Oil Fictions: World Literature and Our Contemporary Petrosphere*. Penn State UP, 2021.

Barad, Karen. "Nature's Queer Performativity." *Kvinder, Køn & Forskning* 1–2, 2012. https:// doi.org/10.7146/kkf.v0i1-2.28067.

Bardolph, Jacqueline. "Ken Saro-Wiwa's *Basi and Company*: Voices from Folk Tale to TV Comedy to Short Stories." *Commonwealth* 19.1 (Fall 1996): 16–23.

Baker, Aryn. "Kano's Jane Austen." *Time*. 17 May 2018.

Barnett, Cynthia. *The Sound of the Sea: Seashells and the Fate of the Oceans*. W. W. Norton, 2021.

Barnwell, Andrea D. "Spirits in Steel: The Art of the Kalabari Masquerade African Arts." *African Arts*. 31. 4 (1998): 80–82.

Baron, Jaimie. "Reframing the Perpetrator's Gaze," *Reuse, Misuse, Abuse: The Ethics of Audiovisual Appropriation in the Digital Era*. Rutgers UP, 2021: 124–53.

Bassey, Nnimmo. "A Living Memorial for Deadened Memories." *EnviroNews Nigeria*. 4 Feb. 2016. Web. 31 Jan. 2025. https://www.environewsnigeria.com/living-memorial -deadened-memories/.

Becker, Becky. "Nollywood: Film and Home Video, or the Death of Nigerian Theatre." *Theatre Symposium* 19.1 (2011): 69–80.

Berends, Andrew. "Andrew Berends on Why He Made His Documentary 'Delta Boys.'" *Sundance Institute*. 2 Dec. 2011. Web. 4 Oct. 2024. https://www.sundance.org/blogs /creative-distribution-initiative/kickstart-delta-boys/.

Black November (*BN*). Dir. Jeta Amata. Wells & Jeta Entertainment. 2012. Film.

"Black November." *Nollywood Reinvented*, 11 July 2014. Web. 7 Jan. 2024. https://www.nollywoodreinvented.com/2014/07/black-november.html.

Blake, William. *Europe Supported by Africa and America*. 1796. Victoria and Albert Museum, London, 2023. Web. 11 Oct. 2023. https://collections.vam.ac.uk/item/O127397/europe-supported-by-africa-and-print-william-blake/.

Blood & Oil (*B&O*) (*Oloibiri*). Dir. Curtis Graham. Netflix. 2015. Film.

"Blood & Oil." *Nollywood Reinvented*, 1 Oct. 2019. Web. 21 Jan. 2024. https://www.nollywoodreinvented.com/2019/10/oloibiri.html.

Bojang, Sheriff, Jr. "UK Court of Appeal rules in favour of two Nigerian communities against Shell." The Africa Report. 14 Oct 2024. https://www.theafricareport.com/364577/uk-court-of-appeal-rules-in-favour-of-two-nigerian-communities-against-shell/. Accessed 28 Jan 2025.

Botting, Fred. "Relations of the Real in Lacan, Bataille and Blanchot." *SubStance* 23.1.73 (1994): 24–40.

Boyer, Dominic. "Infrastructure, Potential Energy, Revolution." *The Promise of Infrastructure*. Eds. Nikhil Anand, Akhil Gupta, and Hannah Appel. Duke UP, 2018. 223–44.

Boyle, Shane [@brechtfast]. "Reminder: when Brecht says that 'Petroleum balks at. . . .'" X. 8 Aug. 2022. Web. 12 Dec. 2024. https://x.com/brechtfast/status/1556987557740777474.

Brecht, Bertolt. *Brecht on Theatre: The Development of an Aesthetic*. Ed. John Willett. Farrar, Straus and Giroux, 1964.

———. "The Street Scene: A Basic Model for an Epic Theatre." 1938. Web. 21 Feb. 2024. https://sites.evergreen.edu/arunchandra/wp-content/uploads/sites/395/2020/08/streetScene.pdf.

"Britain Gets Nigerian Oil." *Vancouver Sun*. 29 Mar. 1958, 17.

Brockes, Emma. "Chimamanda Ngozi Adichie: 'Don't We All Write About Love? When Men Do It, It's a Political Comment. When Women Do It, It's Just a Love Story.'" *The Guardian*. 22 Mar. 2014. Web. 31 Jan. 2025. https://www.theguardian.com/books/2014/mar/21/chimamanda-ngozi-adichie-interview.

Brodsky, Judith. "Kappy and Stella." *Rhode Island Jewish Historical Association Notes* 14.2 (Nov. 2004): 308.

Brooke, James. "Enugu Journal: 30 Million Nigerians Are Laughing at Themselves." *New York Times*, 24 July 1987, *ProQuest*. 5 Mar. 2024.

"bunker, Verb." *Oxford English Dictionary Online*. Mar. 2015. Web. 6 Nov. 2014.

Burkitt, Katharine. *Literary Form as Postcolonial Critique*. Ashgate, 2012.

Business Wire via The Motley Fool. "MoneyGram Warns Consumers That Scammers Will Try to Steal Their Money Along With Their Hearts on Valentine's Day: How to Stop Scammers in Their Tracks on the Road to Romance." Yahoo! News. 22 Jan. 2013. Web. 1 Feb. 2025. https://www.yahoo.com/news/2013-01-22-moneygram-warns-consumers-that-scammers-will-try-t.html.

Cabrera, Christine, and Amy Dana Ménard. "'She Exploded into a Million Pieces': A Qualitative and Quantitative Analysis of Orgasms in Contemporary Romance Novels." *Sexuality & Culture* 17.2 (2013): 193–212.

Campbell, John. "Kidnapping in Nigeria: A Growth Industry." Council on Foreign Relations. 29 May 2020. https://www.cfr.org/blog/kidnapping-nigeria-growth-industry. Accessed 29 Ap 2025.

Canjels, Rudmer. "Creating Partners in Progress: Shell Communicating Oil during Nigeria's Independence." *Petrocinema: Sponsored Film and the Oil Industry*, edited by Marina Dahlquist and Patrick Vonderau. Bloomsbury, 2021: 208–24.

Capretti, Elena. *Botticelli*. Giunti, 1997.

Carswell, Joshua. "Petrohorror and Unknowing: Petrocultural Engagements with the Limits of Philosophical Thought," Parts 1 and 2. *orbistertius*. Nov. 2017.

"The Characters of Botticelli's *Primavera*." Virtual Uffizi: The Unofficial Guide to the Uffizi. 11 Oct. 2023. Web. 31 Jan. 2025. https://www.virtualuffizi.com/the-characters-of -botticelli%E2%80%99s-primavera-.html.

Chijoke, Arinze. "Nigeria Moves to Restore Mangrove Ecosystems in Niger Delta." *International Policy Digest*. 19 May 2021. Web. 20 Mar. 2024. https://intpolicydigest.org /nigeria-moves-to-restore-mangrove-ecosystems-in-niger-delta/.

Chovwen, Anthony. "Delta State Labour Market Assessment Report Conducted Under the Niger Delta Youth Employment Pathways Project, Conducted by an Independent Assessor, December 2020." PIND Foundation Nigeria, 2021.

Cioffi, Sandy. Director's Statement. *Sweet Crude* Press Kit.

Clark, John Pepper. *All for Oil*. Alexander Street Press, 2000.

———. *The Wives' Revolt*. Alexander Street Press, 1985.

Cohn, Jan. *Romance and the Erotics of Property: Mass-Market Fiction for Women*. Duke UP, 1988.

Cole, Jennifer. "The Failure of Polio Eradication: Blame Geopolitics, Not Religion." *Georgetown Journal of International Affairs*. 16 Apr. 2015. Web. 29 Feb. 2024. https://gjia.georgetown .edu/2015/04/16/the-failure-of-polio-eradication-blame-geopolitics-not-religion/.

Cole, Teju. "The White-Savior Industrial Complex." *The Atlantic*. Mar 21, 2012.

Comaroff, Jean and John L. "Alien-Nation: Zombies, Immigrants, and Millennial Capitalism," *The South Atlantic Quarterly* 10.4 (2002): 779–805.

Cooper, Ann. *Maelstrom*. Harlequin Books, 1984.

Cordaid. n.d. Web. 4 Oct. 2024. https://www.cordaid.org/en/.

Cranny-Francis, Anne. *Feminist Fiction: Feminist Uses of Generic Fiction*. St Martin's Press, 1990.

Daggett, Cara. "Petro-Masculinity: Fossil Fuels and Authoritarian Desire," *Millennium: Journal of International Studies* 47.1 (2018): 25–44.

———. "Petro-Masculinity and the Politics of Climate Refusal." Autonomy Institute. 1 May 2022. Web. 31 Jan. 2025. https://autonomy.work/portfolio/petro-masculinity-climate -refusal/.

Dahlquist, Marina, and Patrick Vonderau, eds. *Petrocinema: Sponsored Film and the Oil Industry*. Bloomsbury, 2021.

Damluji, Mona. "The Image World of Middle Eastern Oil." *Subterranean Estates: Life Worlds of Oil and Gas*, eds. Hannah Appel, Arthur Mason, and Michael Watts. Cornell UP, 2015. 147–64.

Daughters of the Niger Delta. Dir. Ilse Van Lamoen. Abuja, Nigeria: Media Information Narrative Development (MIND); Nijmegen, the Netherlands: FLL. 2012. Film.

Davis, Emily S. *Rethinking the Romance Genre: Global Intimacies in Contemporary Literary and Visual Culture*. Palgrave Macmillan, 2013.

Davis, Jack. "Two Against Nature? Brecht, Morton and Contradiction." *Colloquia Germanica*, 53.2/3 (Dec. 2021): 215–32.

Debrix, François. *Global Powers of Horror: Security, Politics, and the Body in Pieces*. Taylor & Francis, 2016.

De Certeau, Michel. *The Practice of Everyday Life*. U of California P, 1988.

Delta Boys. Dir. Andrew Berends. Sundance. 2012. Film.

Derrida, Jacques. "Hospitality." *Angelaki: Journal of Theoretical Humanities* 5.3 (Dec. 2000): 3–18.

Diabate, Naminata. "The Cinematic Language of Naked Protest." *Critical Interventions* 11.3 (Dec. 2017): 248–68.

"documentary, Adjective & Noun." *Oxford English Dictionary Online*, Sept. 2023. Web. 10 Jan. 2024.

Douglas Camp, Sokari. "About Sokari." n.d. Web. 12 Oct. 2023. https://sokari.co.uk/about/.

———. "All That Glitters." 2013. 11 Oct. 2023. https://sokari.co.uk/project/all-that-glitters/.

———. "Big Alagba & Sekibo." 1994. Web. 17 Nov. 2023. https://sokari.co.uk/project/big -alagba-sekibo/.

———. "Blind Love and Grace." 17 Apr. 2017. Web. 11 Oct. 2023. https://sokari.co.uk/?s=blind+love.

———. *Europe Supported by Africa and America.* 2015. Web. 11 Oct. 2023. https://sokari.co.uk/project/europe-supported-by-africa-and-america/.

———. *Green Leaf Barrel.* RDN Arts. 2014. Web. 11 Oct. 2023. https://rdnarts.com/artwork/green-leaf-barrel-2014/.

———. *Ken Saro-Wiwa Living Memorial.* 2006. Web. 11 Oct. 2023. https://sokari.co.uk/project/ken-saro-wiwa-living-memorial/.

———. *My World Your World.* 1997. Web. 30 Oct. 2023. https://sokari.co.uk/project/my-world-your-world/.

———. Personal interview. 6 November 2023.

———. *Primavera.* 2015. Web. 11 Oct. 2023. https://sokari.co.uk/project/primavera/.

———. *Stoa 169.* 2020. Web.4 Nov. 2023. https://sokari.co.uk/project/stoa-169/.

Duncan, Pansy. "Fade to Crude: Petro-Horror and Kubrick's *The Shining.*" *After Kubrick: A Filmmaker's Legacy.* Ed. Jeremi Szaniawski. Bloomsbury, 2020. 179–94.

Eboh, Camillus. "Nigeria's Top Court Says Shell's Appeal Should Be Heard After Oil Spill Claim." *Reuters.* 5 Jan. 2024. Web. 19 Mar. 2024. https://www.reuters.com/world/africa/nigerias-top-court-says-shells-appeal-should-be-heard-after-oil-spill-claim-2024-01-05/.

Edgar, David. "Enter the New Wave of Political Playwrights." *The Guardian.* 27 Feb. 2010. Web. 19 Feb. 2024. https://www.theguardian.com/stage/2010/feb/28/david-edgar-new-political-theatre.

Edwin, Shirin. *Privately Empowered: Expressing Feminism in Islam in Northern Nigerian Fiction.* Northwestern UP, 2016.

Eghagha, Hope. *The Oily Marriage.* Malthouse Press, 2018.

Egya, S. E. "The Pristine Past, the Plundered Present: Nature as Lost Home in Tanure Ojaide's Poetry." *Journal of Commonwealth Literature,* 56.2 (July 2018): 186–200.

———. "Re: your insight into petrocriticism." Email received by Helen Kapstein. 2 Sept. 2024.

Eke, Maureen N. "Introduction: Who Shall Silence the Drums? Another Women's War." *What Mama Said: An Epic Drama,* Osonye Tess Onwueme. Wayne State UP, 2003: 7–12.

Ekpo, Okon. "Movie Review: Oloibiri Is Not the Definitive Niger Delta Narrative." *Ynaija.com.* 28 Oct. 2016. Web. 11 Jan. 2024. https://ynaija.com/movie-review-oloibiri-not-definitive-niger-delta-narrative/.

Elam, Diane. *Romancing the Postmodern.* Routledge, 1993.

Eshun, Kodwo. "Further Considerations on Afrofuturism." *CR: The New Centennial Review* 3.2 (Summer 2003): 287–302.

Esty, Joshua D. "Excremental Postcolonialism." *Contemporary Literature* 40.1 (Spring 1999): 22–59.

"Europe Supported by Africa and America by Sokari Douglas Camp," Victoria and Albert Museum, London, 2023. Web. 11 Oct. 2023. https://www.vam.ac.uk/articles/europe-supported-by-africa-and-america-by-sokari-douglas-camp/.

Evers, Shoshanna. "The Secret Formula of Most Romance Novels." *The Writer's Challenge.* 23. Sept. 2009. Web. June 25, 2018. http://www.thewriterschallenge.com/2009/09/secret-formula-of-most-romance-novels.html.

"fabric, Noun." *Oxford English Dictionary Online.* n.d. Web. Sept. 2023.

Faccini, Dominic et al. "Critical Dictionary." *October* 60 (Spring 1992): 25–31.

Farrell, Riley. "The Oil-Soaked Bird That Shocked the World." BBC. 4 Oct. 2023. https://www.bbc.com/future/article/20231002-the-photo-of-the-deepwater-horizon-bird-that-shocked-the-world Accessed 29 Jan 2025.

Faucon, Benoit. "Nigerian Oil Thieves' Toll." *The Wall Street Journal.* 4 Mar 2009.

Federici, Silvia. *Revolution at Point Zero: Housework, Reproduction, and Feminist Struggle.* PM Press, 2012.

Flitter, Emily. "JPMorgan's Role in Nigerian Oil Deal Has Come Back to Haunt It." *The New York Times* 28 March 2019. Web. 12 Dec. 2024. https://www.nytimes.com/2019/03/28/business/jpmorgan-nigeria.html.

Foote, Stephanie. "Introduction to Stuckness." *Stuckness*, special issue of *Post45*. 19 Sept. 2023. Web. 12 Dec. 2024. https://post45.org/2023/09/introduction-to-stuckness/.

Forbes Woman Africa, Dec 2013–Jan 2014. Cover.

"Four in Seattle-Based Film Group Arrested in Nigeria," *The Seattle Times*. 14 Apr. 2008. https://www.seattletimes.com/seattle-news/four-in-seattle-based-film-group-arrested-in-nigeria/ Accessed 30 Jan 2025.

Freestone, Elizabeth, and Jeanie O'Hare. *100 Plays to Save the World*. Theatre Communications Group, 2023.

Fu, Xiangyi. "Horror Movie Aesthetics: How Color, Time, Space and Sound Elicit Fear in an Audience." Diss., Northeastern U, 2016.

Ghosh, Amitav. *The Great Derangement: Climate Change and the Unthinkable*. U of Chicago P, 2016.

———. *The Nutmeg's Curse: Parables for a Planet in Crisis*. U of Chicago P, 2022.

———. "Petrofiction." *New Republic* 206.9 (1992): 29–34.

Galal, Saifaddin. "Internet penetration in Africa January 2024, by country." Statista. Mar 14, 2024. Web. 29 Jan 2025. https://www.statista.com/statistics/1124283/internet-penetration-in-africa-by-country/#:~:text=Varying%20but%20growing%20levels%20of,1.1%20billion%20users%20by%202029.

Genova, James. "Film, Radio, and Society in Colonial and Postcolonial Africa," *Oxford Research Encyclopedia of African History*, 19 Dec. 2017, https://doi.org/10.1093/acrefore/9780190277734.013.115.

Gordon, Glenna. *Diagram of the Heart*. Red Hook Editions, 2016.

Gordon, Orin. "Nigeria's Growing Number of Female Oil Bosses." *BBC News*. Business. 11 Sept. 2014.

Goyal, Yogita. "Introduction: Africa and the Black Atlantic." *Research in African Literatures* 45.3 (Fall 2014): v–xxv.

Green-Simms, Lindsey. "Occult Melodramas: Spectral Affect and West African Video-Film," *Camera Obscura: Feminism, Culture, and Media Studies* 27.2.80 (1 Sept. 2012): 25–59.

———. *Postcolonial Automobility: Car Culture in West Africa*. U of Minnesota P, 2017.

Griswold, Wendy. *Bearing Witness: Readers, Writers, and the Novel in Nigeria*. Princeton UP, 2000.

GSMA. *The Mobile Economy 2014*. London, 2014.

Halberstam, Judith [Jack]. *Skin Shows: Gothic Horror and the Technology of Monsters*. Duke UP, 1995.

Hall, Stuart. "The Local and the Global: Globalization and Ethnicity." *Dangerous Liaisons: Gender, Nation, and Postcolonial Perspectives*. Ann McClintock et al. U of Minnesota P, 1997. 173–87.

Hallemeier, Katherine. "'To Be from the Country of People Who Gave': National Allegory and the United States of Adichie's *Americanah*." *Studies in the Novel* 47.2 (Summer 2015): 231–45.

Hammer, Joshua. "World's Worst Traffic Jam." *The Atlantic*, July/Aug. 2012.

Haraway, Donna. "Teddy Bear Patriarchy: Taxidermy in the Garden of Eden, New York City, 1908–1936." *Social Text* 11 (Winter 1984): 20–64.

Haynes, Jonathan. "'New Nollywood': Kunle Afolayan," *Black Camera* 5.2 (Spring 2014): 53–73.

———. *Nigerian Video Films*. University Center for International Studies, 2000.

"A Head for Business and Fists for a Fight." *Forbes Africa*. 1 Aug. 2016.

Hetherington, Kregg. "Surveying the Future Perfect: Anthropology, Development and the Promise of Infrastructure." *Infrastructures and Social Complexity: A Companion*, eds. Penelope Harvey, Casper Bruun Jensen, and Atsuro Morita. Routledge, 2017. 40–50.

Hitchcock, Peter. *The Long Space: Transnationalism and Postcolonial Form.* Stanford UP: 2010.
———. "Oil in an American Imaginary." *New Formations* 69.4 (2010): 81–97.
Hodapp, James. "A Serious Television Trickster: Ken Saro-Wiwa's Political and Artistic Legacy in *Basi and Company.*" *Journal of Postcolonial Writing* 54:4 (2018): 504–14.
"hostage, Noun[1]." *Oxford English Dictionary Online*, Sept. 2023. Web.
Houghton, Gerard. "The Fabric of the World" ("FW"). *Jonkannu Masquerade.* London: October Gallery, 1 June 2022.
———. "Sokari Douglas Camp C.B.E.: A New Allegory for Spring" ("C.B.E.") *Primavera.* London: October Gallery, 2016.
Hutchings, Kevin. *Romantic Ecologies and Colonial Cultures in the British Atlantic World, 1770–1850,* McGill-Queen's UP, 2009.
Ifedigbo, Sylva Nze. "Have [Nigerian] Romance Novels Come of Age?" *Critical Literature Review,* 24 Jan. 2010. Web. 12 Dec. 2024. https://criticalliteraturereview.blogspot.com /2010/01/have-nigerian-romance-novels-come-of.html.
"Imperiled in Pregnancy." TVTropes. 4 Oct 2024. Web. 12 Dec. 2024. https://tvtropes.org /pmwiki/pmwiki.php/Main/ImperiledInPregnancy.
Ingram, David. *Green Screen: Environmentalism and Hollywood Cinema.* U of Exeter P, 2000.
Iwowo, Samantha. "The Problematic Mise En Scène of NeoNollywood." *Communication Cultures in Africa* 2.1 (June 2020): 93–119.
Izuzu, Chimubga. "'Oloibiri' Creatively Encapsulates the Plight of the Niger Delta Region." *Pulse.* 23 Oct 2016. Web. 11 Jan. 2024. https://www.pulse.ng/entertainment/movies /pulse-movie-review-oloibiri-creatively-encapsulates-the-plight-of-the-niger-delta /zeqlbng.
Jameson, Fredric. *Postmodernism or, The Cultural Logic of Late Capitalism.* Duke UP, 1995.
James, Sienna. "Sokari Douglas Camp on Steel Sculptures, the Welding Process and Slave Legacies." *Voice.* 9 Oct. 2023. Web. 16 Nov. 2023. https://www.voicemag.uk/interview /12985/sculptor-sokari-douglas-camp-on-steel-welding-and-slave-legacies.
Johnston, Ian. "Campaign to Put Ecocide on a Par with Genocide in an Attempt to Curb Environmental Destruction." *The Independent* (UK). 21 Oct. 2014. Web. 26 Mar. 2015. https://www.the-independent.com/climate-change/news/campaign-to-put-ecocide-on-a -par-with-genocide-in-attempt-to-curb-environmental-destruction-9789297.html.
Jones, Ann Rosalind. "Mills & Boon Meets Feminism." *The Progress of Romance: The Politics of Popular Fiction,* ed. Jean Radford. Routledge, 1986. 195–218.
Jones, Christopher F. *Routes of Power: Energy and Modern America.* Harvard UP, 2014.
Habila, Helon. *Oil on Water.* W. W. Norton: 2010.
Horton, Robin. "The Kalabari 'Ekine' Society: A Borderland of Religion and Art." *Africa: Journal of the International African Institute* 33.2 (1963): 94–114.
Kadafa, Adati Ayuba. "Environmental Impacts of Oil Exploration and Exploitation in the Niger Delta of Nigeria." *Global Journal of Science Frontier Research Environment & Earth Sciences* 12.3 (2012). Web. 21 Jan. 2024. http://large.stanford.edu/courses/2017/ph240 /nwagbo1/docs/kadafa.pdf.
Kapstein, Helen, *Postcolonial Nations, Islands, and Tourism: Reading Real and Imagined Spaces.* Rowman & Littlefield Publishers, 2019.
Kapstein, I. J. "The Meaning of Shelley's 'Mont Blanc.'" *Publications of the Modern Language Association of America* 62.4 (1947): 1046–60.
Kerridge, Richard. "Ecocritical Approaches to Literary Form and Genre: Urgency, Depth, Provisionality, Temporality." *The Oxford Handbook of Ecocriticism,* ed. Greg Garrard. Oxford UP, 2014. 361–76.
Khairy, Wael. "Why The Thing Is One of the Most Effective Horror Movies Ever Made." RogerEbert.com. 26 Oct. 2020. Web. 3 Oct. 2024. https://www.rogerebert.com/far -flung-correspondents/why-the-thing-is-one-of-the-most-effective-horror-movies-ever -made.

Khanna, Rakesh. "A Note from the Publishers." Yakubu, Balaraba Ramat. *Sin is a Puppy That Follows You Home*. First published 1990. Translated by Aliyu Kamal. Chennai, India: Blaft. 2012: v–vi.

Koerner, Brendan. "Online Dating Made This Woman a Pawn in a Global Crime Plot." *Wired*. 5 Oct. 2015.

Kornbluh, Anna. *The Order of Forms: Realism, Formalism, and Social Space*. U of Chicago P, 2019.

Krimsky, Sheldon. "The Funding Effect in Science and Its Implications for the Judiciary," *Journal of Law and Policy* 13.1 (Jan. 2005): 43–68.

Kuforiji, Remi. "Water No Get Enemy." 24 Nov. 2023. Webinar.

Labban, Mazen. "Oil Spill: Inside the Global Market for Crude." *The American Prospect*. 28 Apr. 2020. Web. 19 Mar. 2024. https://prospect.org/environment/absurd-global-market-crude-negative-oil-futures/.

Larkin, Brian. "Promising Forms: The Political Aesthetics of Infrastructure." *The Promise of Infrastructure*. Eds. Nikhil Anand, Akhil Gupta, and Hannah Appel. Duke UP, 2018. 175–202.

———. *Signal and Noise: Media, Infrastructure, and Urban Culture in Nigeria*. Duke UP, 2008.

Leetsch, Jennifer. "Love, Limb-Loosener: Encounters in Chimamanda Adichie's *Americanah*." *Journal of Popular Romance Studies* 6 (12 Apr. 2017). Web. 23 June 2018. https://www.jprstudies.org/2017/04/love-limb-loosener-encounters-in-chimamanda-adichies-americanahby-jennifer-leetsch/.

LeMenager, Stephanie. "The Aesthetics of Petroleum, After *Oil!*" *American Literary History* 24.1 (Spring 2012): 59–86.

———. "Eden If We Dare." *Zina Saro-Wiwa: Did You Know We Taught Them How to Dance?* Ed. Maria Bailey. Blaffer Art Museum, 2016. 39–45.

———. *Living Oil: Petroleum Culture in the American Century*. Oxford UP: 2014.

Levine, Caroline. *Forms: Whole, Rhythm, Hierarchy, Network*. Princeton UP, 2017.

Lincoln, Sarah. "Rotten English: Excremental Politics and Literary Witnessing." *Encountering the Nigerian State: Excess and Abjection*, eds. Wale Adebanwi and Ebenezer Odabare. Palgrave-Macmillan, 2010: 79–98.

Lindsay, James. "Queer (v)." *New Discourses*. 4 June 2020. Web. 14 Feb. 2024. https://newdiscourses.com/tftw-queer-v/.

"The Living Memorial." *Platform*. n.d. Web. 4 Nov. 2023. https://platformlondon.org/background-2/.

Longley, James. "Reflections on Andrew Berends," International Documentary Association, 8 Mar. 2019, Web, 4 Apr. 2024, https://www.documentary.org/online-feature/reflections-andrew-berends.

Macdonald, Graeme, ed. "Commentary." *The Cheviot, the Stage, and the Black, Black Oil*, John McGrath. Bloomsbury, 2015. 17–80.

———. "Improbability Drives: The Energy of SF." *Paradoxa* 26 (2014): 111–44.

———. "Oil and World Literature." *American Book Review* 33.3 (Mar.–Apr. 2012): 7–31.

———. "Research Note: The Resources of Fiction." *Reviews in Cultural Theory* 4.2 (2013): 1–24.

Makdisi, Saree. *William Blake and the Impossible History of the 1790s*. U of Chicago P, 2002.

Malm, Andreas. *Fossil Capital: The Rise of Steam Power and the Roots of Global Warming*. Verso, 2016.

Matthews, Chris. "Will e-Publishing Help Africa to Switch on to Reading?" *BBC News*. 2 Dec. 2013. Web. 25 Mar. 2015.

Mbembe, Achille. *Necropolitics*. Duke UP: 2019.

———. *Out of the Dark Night: Essays on Decolonization*. Columbia UP: 2021.

McDonald, Grantley. "Georgius Agricola and the Invention of Petroleum." *Bibliothèque d'Humanisme et Renaissance* 73.2 (2011): 351–63.

McGillis, Roderick. "Postcolonialism and Recovery: A Future Evermore About to Be." *Postcolonial Theory in the Global Age: Interdisciplinary Essays*. Eds. Om Prakash Dwivedi and Martin Kich. McFarland, 2013. 21–34.

McMahon, Barbara. *Marrying the Scarred Sheikh*. Harlequin, 2010.

"mechanic, Adjective & Noun." *Oxford English Dictionary Online*. Mar. 2015. Web. 6 Nov. 2014.

Mellor, Anne K. "Sex, Violence, and Slavery: Blake and Wollstonecraft." *Huntington Library Quarterly* 58.3/4 (1995): 345–70.

Milton, John. *Areopagitica*. 1644. Project Gutenberg. 23 Feb. 2013. Web. 12 Dec. 2024. https://www.gutenberg.org/cache/epub/608/pg608-images.html.

Moore, Sophie Sapp, et al. "Plantation Legacies." *Edge Effects*. 15 May 2021. Web. 27 Mar. 2024. https://edgeeffects.net/plantation-legacies-plantationocene/.

Murdock, Heather. "Nigeria's Nollywood: Most Prolific Movie Machine." *AP*. 3 Apr. 2014, Web. 12 Jan. 2024. https://apnews.com/1bf5d18c1d0c470aa721ecea8fdaa6f9.

The Naked Option: A Last Resort. Dir. Schermerhorn, Candace. Filmmakers Library, 2011. Film.

Nader, Ralph. "Sweet Crude Buzz." 2007. Web. 12 Dec. 2024. http://www.sweetcrudemovie.com/.

Nesbith, N. Graham. Untitled Review of *What Mama Said: An Epic Drama* by Osonye Tess Onweume. *African American Review*. 2005 39.1–2 (Spring–Summer 2005): 256–57.

"Nigeria: Hundreds of Oil Spills Continue to Blight Niger Delta." Amnesty International. 19 Mar. 2015. Web. 25 Mar. 2015. https://www.amnesty.org/en/latest/news/2015/03/hundreds-of-oil-spills-continue-to-blight-niger-delta/.

"Nigeria Is a Bookless Country—Prof. Emenanjo." *Vanguard*. 7 July 2013. Web. 21 Mar. 2015. https://www.vanguardngr.com/2013/07/nigeria-is-a-bookless-country-prof-emenanjo/.

"Nigeria: Potential, Growth and Challenges." Shell Global. n.d. Web. 25 Mar. 2015. Defunct.

"Nigerian Oil Strike Made, Company Says." *The Montreal Gazette*. 18 Aug. 1956, 2.

Nixon, Rob. *Slow Violence and the Environmentalism of the Poor*. Harvard UP, 2011.

Noor, Dharna. "Fossil Fuel Firms Could Be Tried in US for Homicide over Climate-Related Deaths, Experts Say." *The Guardian*. 21 Mar. 2024. Web. 21 Mar. 2024. https://www.theguardian.com/us-news/2024/mar/21/fossil-fuel-companies-homicide-climate-deaths-lawsuit.

"Nothing Is Scarier." TVTropes. n.d. Web. 5 Oct 2024. https://tvtropes.org/pmwiki/pmwiki.php/Main/NothingIsScarier.

Nwakunor, Gregory Austin. "Oloibiri Story Is Evergreen, Will Remain Relevant for a Long Time," *The Guardian* (Nigeria). 9 May 2021. Web. 17 Jan. 2024. https://guardian.ng/sunday-magazine/oloibiri-story-is-evergreen-will-remain-relevant-for-a-long-time/.

Obafemi, Frances N, Uchechi R Ogbuagu, and Emmanuel Nathan. "Petroleum Resource, Institutions, and Economic Growth in Nigeria," *Journal of Business & Management* 1.3 (2013): 154–65.

Odufuwa, Damilola. "Nigeria: The Women Who Reject Feminism." CNN. 8 Oct. 2008. https://www.cnn.com/2018/10/08/africa/nigeria-gender-wars-twitter-feminism/index.html Accessed 26 Jan 2025.

"Ogoni Bill of Rights: 31 Years After and Still No Justice." *IMPSDL*. 4 Oct. 2021. Web. 6 Apr. 2024. https://www.ipmsdl.org/statement/ogoni-bill-of-rights-31-years-after-and-still-no-justice/.

"Oil Theft: Shell Divests From Nigeria, Sells Off $737m Upstream Oil Assets." The Street Journal. 20 Mar. 2015. Web. 1 Feb. 2025. https://thestreetjournal.org/oil-theft-shell-divests-from-nigeria-sells-off-737m-upstream-oil-assets/.

Ojaide, Tanure. *Delta Blues and Home Songs: Poems*. Kraft Books, 1998.

Okorafor, Nnedi, and Foster, Alan Dean. *Kabu Kabu*. United States, Prime Books, 2013.

Okorafor, Nnedi. "The Popular Mechanic." *InterNova: International Science Fiction*. Ed. Michael Iwoleit. 2011. Web. 5 Jan. 2012. Defunct.

———. "Spider the Artist." *Kabu-Kabu*. Prime Books, 2013, 101–15.

Okoro, Emmanuel Emeka et al. "Radiological and Toxicity Risk Exposures of Oil Based Mud: Health Implication on Drilling Crew in Niger Delta," *Environmental Science and Pollution Research* 27 (2020): 5387–97.

Okpadah, Stephen Ogheneruro. "Politics, Oil, and Theater in Africa." *Environmental Postcolonialism: A Literary Response.*Eds. Shubhanku Kochar and M. Anjum Khan. Lexington Books, 2021. 47–60.

Okukpon, Irekpitan. Syllabus for Advanced Oil & Gas Law I, National Open University of Nigeria, Abuja, 2022. https://nou.edu.ng/coursewarecontent/PUL%20821.pdf.

Okunoye, Oyeniyi. "Alterity, Marginality and the National Question in the Poetry of the Niger Delta," *Cahiers d'études africaines,* 48.191 (2008): 413–36.

Olaoluwa, Senayon. "Dislocating Anthropocene: The City and Oil in Helon Habila's *Oil on Water.*" *ISLE: Interdisciplinary Studies in Literature and Environment* 27.2 (2020): 243–67.

Onuah, Felix. "Nigerian State-Owned Company Starts Oil Drilling at Field in North." *Reuters.* 4 Nov. 2022. Web. 4 Apr. 2024. https://www.reuters.com/business/energy/state-owned -company-starts-oil-drilling-field-northern-nigeria-2022-11-22/.

Onwuegbuchi, Chike. "Nigeria, Others Emerge Highest Mobile Online Reading Population, Says Report." *The Guardian.* 14 Sept. 2018. Web. 22 Aug. 2019.

Onwueme, Osonye Tess. *Then She Said It.* Alexander Street Press, 2001.

———. *What Mama Said: An Epic Drama,* Wayne State UP, 2003.

Orr, David. "Nigerian Minorities Claim Oil Wealth as Stolen Birthright." *Independent* 3 Jan. 1996. Web. 19 Mar. 2024. https://www.independent.co.uk/news/world/nigerian -minorities-claim-oil-wealth-as-stolen-birthright-1322173.html.

Over, William. "Redefining Political Drama: Onwueme and Nigerian Society." *Contemporary Justice Review* 13:2 (2010): 173–89.

Oyibo, Chief. "Shell upbeat at Nigeria oil spill ruling." Oyibos Online. 12 Mar. 2015. Web. 1 Feb. 2025. http://www.oyibosonline.com/shell-upbeat-at-nigeria-oil-spill-ruling/.

Persson. Jennie. "Increasing Economic Opportunity for Residents in the Niger Delta: A Problem-Driven Political Economy Analysis." *International Affairs Review,* 15 June 2018. Web. 4 Feb. 2025. https://www.iar-gwu.org/print-archive/fj8yy35dkvsh543 lfuslvxionpq5q0.

Petrocultures Research Group, *After Oil.* Petrocultures Research Group, 2016.

Price, Jennifer. "A Brief Natural History of the Plastic Pink Flamingo." Chap. 3 in *Flight Maps: Adventures With Nature In Modern America.* New York: Basic Books, 2000. (Adaptation for The American Scholar) http://www.environmentandsociety.org/node/6448.

Poe, Edgar Allan. "Twice-Told Tales: A Review." *Graham's Magazine.* May 1842.

Puri, Jyoti. "Reading Romance Novels in Postcolonial India." *Gender and Society* 11.4 (Aug. 1997): 434–52.

Rabin, Nathan. "Black November." *The Dissolve,* 8 Jan. 2015. Web. 15 Jan 2024. https:// thedissolve.com/reviews/1302-black-november/.

Radway, Janice A. *Reading the Romance: Women, Patriarchy, and Popular Literature.* U of North Carolina, 1984.

"recovery, Noun." *Oxford English Dictionary Online.* Mar. 2024. Web. 12 Dec. 2024.

Reddick, Yvonne. "Palm Oil and Crude Oil: Environmental Damage, Resource Conflict, and Literary Strategies in the Niger Delta," *ISLE: Interdisciplinary Studies in Literature and Environment,* 26.3 (Summer 2019): 688–721, https://doi.org/10.1093/isle/isz061.

Regis, Pamela. *A Natural History of the Romance Novel.* U of Penn. P, 2013.

Rogers, Nicole. "Wilding." *Research Handbook on Law and Literature.* Eds. Peter Goodrich, Daniela Gandorfer, and Cecillia Gebruers. Edward Elgar Publishing, 2022.

"romance, Noun and adjective[1]." *Oxford English Dictionary Online.* June 2017. Web. 26 Nov. 2017.

"saboteur, Noun." *Oxford English Dictionary Online.* Dec. 2023. Web. 12 Dec. 2024.

"sabotage, Noun." *Oxford English Dictionary Online.* Mar. 2015. Web. 6 Nov. 2014.

Saed, Alejandro Hirsh. "Storying the Slow Violence of Environmental Degradation in the Niger Delta." *PopMatters.* 5 Dec. 2022. Web. 3 Jan. 2024. https://www.popmatters.com /storytelling-slow-violence-niger-delta.

Said, Edward. *Culture and Imperialism*. Knopf Doubleday Publishing Group, 2012.

Salmon, C. "The Pop Culture of Sex: An Evolutionary Window on the Worlds of Pornography and Romance." *Review of General Psychology* 16.2 (2012): 152–60.

Saro-Wiwa, Ken. *Basi and Company: A Modern African Folktale*. Saros International, 1987.

———. *Forest of Flowers*. Longman, 1997.

———. *Genocide in Nigeria: The Ogoni Tragedy*. Saros International Publishers, 1992.

———. *A Month and a Day: A Detention Diary*. Penguin Books, 1995.

———. *Sozaboy*. Pearson College Div; First Edition Thus, 1994.

Saro-Wiwa, Zina. *Niger Delta: A Documentary*. 2015. Film. https://www.zinasarowiwa.com/artworks/niger-delta-a-documentary.

Sasu, Doris Dokua. "Internet usage in Nigeria - statistics & facts." Statista. 28 June 2024. https://www.statista.com/topics/7199/internet-usage-in-nigeria/#topicOverview Accessed 25 Jan 2025.

Saunders, William. "The Symbolism of the Pelican." *Arlington Catholic Herald*. 2023. Web. 15 Nov. 2023. https://www.catholiceducation.org/en/culture/catholic-contributions/the-symbolism-of-the-pelican.html.

Scarsini, Valentina. "*Americanah* or Various Observations About Gender, Sexuality and Migration: A Study of Chimamanda Ngozi Adichie." MA Thesis, Universita Ca'Foscari Venezia, 2016–17. Web. 23 June 2018.

Schlote, Christiane. "Oil, Masquerades, and Memory: Sokari Douglas Camp's Memorial of Ken Saro–Wiwa". *Engaging with Literature of Commitment. Volume 1*. Leiden, The Netherlands: Brill, 2012: 241–61.

Schuessler, Jennifer. "What Is a Teddy Roosevelt Presidential Library Doing in North Dakota?" *The New York Times*. 27 Oct. 2023. Web. 30 Oct. 2023. https://www.nytimes.com/2023/10/27/arts/theodore-roosevelt-presidential-library.html.

Schuppli, Susan. "Dirty Pictures." *Living Earth: Field Notes from the Dark Ecology Project 2014–2016*. Sonic Acts Press, 2016. 190–209.

Schuster, Joshua. "Where Is the Oil in Modernism?" *Petrocultures: Oil, Politics, Culture*. Eds. Sheena Wilson, Adam Carlson, and Imre Szeman. McGill-Queen's UP, 2017. 197–213.

Scott, Gordon. "What Is a Naked Option? How Naked Calls and Puts Work." *Investopedia*. 21 Dec 2023. Web. 4 Apr. 2024. https://www.investopedia.com/terms/n/nakedoption.asp.

Sheldon, Rebekah. *The Child to Come: Life After the Human Catastrophe*. U of Minnesota P, 2016.

"Shell in Nigeria: Environmental Performance—Oil Spills." Shell Companies in Nigeria: Shell Petroleum Development Company of Nigeria Limited, Shell Nigeria Exploration and Production Company Limited and Shell Nigeria Gas Limited. Apr. 2013. Web. 25 Mar. 2015. PDF file.

"Shell in Nigeria: Oil Theft, Sabotage and Spills." Shell Petroleum Development Company of Nigeria Limited, Shell Nigeria Exploration and Production Company Limited and Shell Nigeria Gas Limited. Apr. 2014. Web. 25 Mar. 2014. PDF file.

Simpson, Mark and Imre Szeman. "Impasse time." *South Atlantic Quarterly* 120.1 (2021): 77–89.

Smith, Zadie and Chimamanda Ngozi Adichie. "Between the Lines: Chimamanda Ngozi Adichie with Zadie Smith." *Schomburg Center for Research in Black Culture*. 19 Mar. 2014. Web. 24 June 2018.

Snitow, Ann Barr. "Mass Market Romance: Pornography for Women Is Different." *Radical History Review* 20 (1979): 141–61. Rpt. in *Women and Romance: A Reader*. Ed. Susan Ostrov Weisser. NYU Press, 2001. 307–22.

Sofiullahi, Abdulwaheed. "Problems Mount for Sahara Gas Pipeline, Leaving Nigerian Taxpayers at Risk." *Climate Home News*. 14 Feb. 2024. Web. 26 Feb. 2024. https://www.climatechangenews.com/2024/02/14/problems-mount-for-sahara-gas-pipeline-leaving-nigerian-taxpayers-at-risk/.

"Soft-Spoken Sadist." TVTropes. n.d. Web. 3 Oct. 2024. https://tvtropes.org/pmwiki/pmwiki.php/Main/SoftSpokenSadist.

Statista Research Department. "Number of mobile connections in Nigeria 2017–2024." Statista. Jun 28, 2024 https://www.statista.com/statistics/1176097/number-of-mobile -connections-nigeria/#:~:text=As%20of%20January%202024%2C%20Nigeria%20regis- tered%20over,approximately%2091%20percent%20of%20Nigeria's%20total%20popula- tion. Accessed 26 Jan 2025.

Stevens, Quentin, and Karen A. Franck. *Memorials as Spaces of Engagement: Design, Use and Meaning*. Routledge, 2016.

Sweet Crude. Dir. Sandy Cioffi. Cinema Guild. 2009. Film.

Szeman, Imre. "Crude Aesthetics: The Politics of Oil Documentaries." *Journal of American Studies* 46.2 (2012): 423–39.

This Is Nollywood. Dir. Franco Sacchi. California Newsreel. 2007. Film.

THR staff. "'Black November' Film Review." *The Hollywood Reporter*. 7 Jan 2016. Web. 15 Jan 2024https://www.hollywoodreporter.com/movies/movie-reviews/black-november-film -review-761561/.

"Transformation Horror." TVTropes. n.d. Web. 8 Oct. 2024. https://tvtropes.org/pmwiki /pmwiki.php/Main/TransformationHorror.

"The Women of the Niger Delta." The Naked Option's Weblog. n.d. Web. 4 April 2024. https:// nakedopt.wordpress.com/.

Tulsi, Yogesh. "An Oily Mirror: 1950s Orang Minyak Films as Singaporean Petrohorror." *Eating Chilli Crab in the Anthropocene: Environmental Perspectives on Life in Singapore*. Ed. Matthew Schneider-Mayerson. Ethos Books 2020. 159–82.

Turcotte, Heather. "Contextualizing Petro-Sexual Politics." *Alternatives: Global, Local, Political* 36.3 (Aug. 2011): 200–20.

Udechukwu, Obiora. *What the Madman Said: Poems*. Germany, Boomerang Press, 1990.

Ugochukwu, Françoise. "Nollywood and the Niger Delta Conflict—Media in an Advisory Capacity." *AFFRIKA: Journal of Politics, Economics and Society* 8.2 (Dec. 2018): 123–41.

Umez, Uche Peter. "Wild Flames." *Poor Mojo's Almanac(k) Classics 2000–2011*. 16 Feb 2006. Web. 5 Jan 2012. https://www.poormojo.org/cgi-bin/gennie.pl?Fiction+266+bi.

United States. Cong. *Expressing the Sense of Congress That as One of the World's Impor- tant Wetland and Coastal Marine Ecosystems, the Niger Delta Should Be Protected and Its Recovery and Economic Development a Priority*. H. Con. Res. 121 (IH), 112th Cong., 2nd sess. 27 April 2012. Web. 24 Jan 2024. https://www.govinfo.gov/app/details/BILLS -112hconres121ih.

UN News, "Polio Is No Longer Endemic in Nigeria—UN Health Agency," Africa Renewal, Web. 29 Feb. 2024, https://www.un.org/africarenewal/news/polio-no-longer-endemic -nigeria-%E2%80%93-un-health-agency.

Vidal, John. "Niger Delta Oil Spills Clean-Up Will Take 30 Years, Says UN." *The Guardian*. 5 Aug 2011. Web. 20 Mar 2024. https://ourworld.unu.edu/en/niger-delta-oil-spills-clean-up -will-take-30-years-says-un.

"Violence Is Disturbing." TVTropes. n.d. Web. 4 Oct. 2024. https://tvtropes.org/pmwiki /pmwiki.php/Main/ViolenceIsDisturbing.

Vitalicio, Renz. "The 8 Most Common Horror Movie Tropes and Cliches, Explained." WhatNerd.com. 13 May 2022. Web. 4 Oct. 2024. https://whatnerd.com/most-common -horror-movie-tropes-cliches-explained/.

Wainaina, Binyavanga. "How to Write about Africa." *Granta* 92 (2 May 2019): 92–95.

Wainger, Leslie. *Writing a Romance Novel for Dummies*. John Wiley & Sons, 2011.

"Water No Get Enemy." *Econation*. 31 Aug. 2023. Web. 16 Nov. 2023. https://econation.one /blog/water-no-get-enemy/.

Water No Get Enemy: Counter-Cartographies of Diaspora. Dir. Remi Kuforiji. Disembodied Territories. 2023. Web. 1 Nov. 2023. https://disembodiedterritories.com/Water-No-Get -Enemy-Counter-Cartographies-of-Diaspora.

Watts, Laura. *Energy at the End of the World: An Orkney Islands Saga*. MIT Press, 2019.

Watts, Michael J. "Oil as Money: The Devil's Excrement and the Spectacle of Black Gold." *Reading Economic Geography*. Eds. Adam Tickell et al. Wiley, 2008. 205–19.

———. "Petro-Violence: Community, Extraction, and Political Ecology of a Mythic Commodity." *Violent Environments*. Eds. Michael Watts and Nancy Lee Peluso. Cornell UP, 2001. 189–212.

———. "Petro-Violence: Some Thoughts on Community, Extraction, and Political Ecology." UC Berkeley: Institute of International Studies, 1999. *eScholarship*. Web. 20 Mar. 2015. https://escholarship.org/uc/item/7zh116zd.

———. "Sweet and sour: The Curse of Oil in the Niger Delta." *Curse of the Black Gold: 50 Years of Oil in the Niger Delta*, Ed Kashi. PowerHouse Books, 2008. 36–47.

Wenzel, Jennifer. "Behind the Headlines." *American Book Review* 33.3 (Jan 2012): 13–14.

———. *The Disposition of Nature: Environmental Crisis and World Literature*. Fordham UP, 2019.

———. "Petro-Magic-Realism: Toward a Political Ecology of Nigerian Literature." *Postcolonial Studies* 9.4 (Dec 2006): 449–64.

West, Mark and Han Ei Chew, "Reading in the Mobile Era: A Study of Mobile Reading in Developing Countries." Ed. Rebecca Kraut. The United Nations Educational, Scientific and Cultural Organization (UNESCO), 2014.

Whitman, Myne. *A Heart to Mend*. AuthorHouse, 2009.

Whitsitt, Novian. "Islamic-Hausa Feminism Meets Northern Nigerian Romance: The Cautious Rebellion of Bilkisu Funtuwa," *African Studies Review* 46.1 (Apr. 2003): 137–53.

Wilcox, Isobel, "Artist Zina Saro-Wiwa First Solo Show at the Blaffer Museum in Houston." *Happening Africa*. 26 Jan 2016. Web. 13. Dec. 2024. https://www.happeningafrica.com/artist-zina-saro-wiwa-first-solo-show-at-the-blaffer-museum-in-houston/.

Wilkinson, Jane. "'Second-New': Serialization and Circulation in *Basi and Company*." *Telling Stories: Postcolonial Short Fiction in English*. Ed. Jacqueline Bardolph. Rodopi, 2001. 255–67.

Williams, Raymond. *Keywords: A Vocabulary of Culture and Society*. Rev. ed. Fontana, 1983.

Wilson, Sheena. "Gendering Oil: Tracing Western Petrosexual Relations." In *Oil Culture*. Eds. Ross Barrett and Daniel Worden. U of Minnesota P, 2014. 244–66.

Witness History. "Art Fights Oil in Nigeria." *BBC News World Service*. Web. 13 Dec, 2024. https://wspartners.bbc.com/clip/p0836p4n.

Wiwa, Ken. *In the Shadow of a Saint: A Son's Journey to Understand His Father's Legacy*. Steerforth: 2001.

Worden, Daniel. "Fossil-Fuel Futurity: Oil in 'Giant.'" *Journal of American Studies* 46.2, (May 2012): 441–60.

Yaeger, Patricia. "Editor's Column: Literature in the Ages of Wood, Tallow, Coal, Whale Oil, Gasoline, Atomic Power, and Other Energy Sources." *Publications of the Modern Language Association of America* 126.2 (Mar. 2011) 305–26.

Yakubu, Balaraba Ramat. *Sin Is a Puppy That Follows You Home*. Trans. Aliyu Kamal. Blaft, 2012.

The Yes Men, "Shell: We are Sorry." Youtube. 28 Mar 2010. https://www.youtube.com/watch?v=zciWUOrIUqo Accessed 26 Ap 2025.

Young, Allison. "Sokari Douglas Camp." Africanah.org. 12 Aug. 2019. Web. 12 Nov. 2023. https://africanah.org/sokari-douglas-camp-3.

Young, James E. *The Texture of Memory: Holocaust Memorials and Meaning*. Yale UP, 1994.

Yusoff, Kathryn. *A Billion Black Anthropocenes or None*. U of Minnesota P, 2018.

ABOUT THE AUTHOR

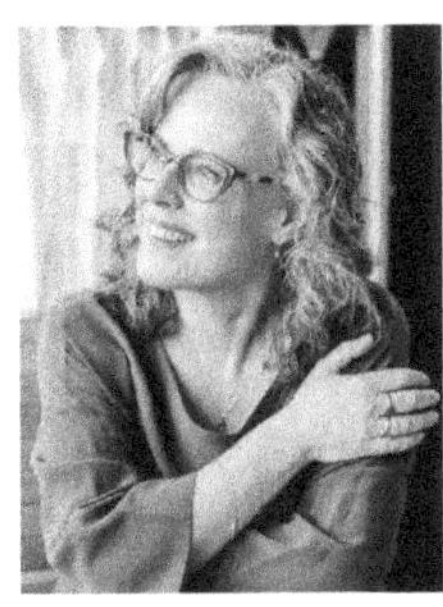 **Helen Kapstein** is a professor in the English Department at John Jay College, The City University of New York. A postcolonial scholar, she earned her PhD in English and Comparative Literature from Columbia University. She is the author of *Postcolonial Nations, Islands, and Tourism: Reading Real and Imagined Spaces.*

www.ingramcontent.com/pod-product-compliance
Lightning Source LLC
Chambersburg PA
CBHW060749240726
48664CB00009BA/1671